The MEDIUM Disclosure II - Constitutive Physics of the Absolute Medium

smbarua

Published by smbarua, 2026.

THE MEDIUM DISCLOSURE II - CONSTITUTIVE PHYSICS OF THE ABSOLUTE MEDIUM

First edition. February 15, 2026.

ISBN: 979-8233933509

Written by smbarua.

To

The Visionaries

The MEDIUM Disclosure II

CONSTITUTIVE PHYSICS OF THE ABSOLUTE MEDIUM

Preface

Preface

For more than a century, physics has advanced by embracing abstractions: spacetime without substance, fields without medium, quantum probabilities without mechanism. Yet beneath these powerful mathematical tools lies an unresolved tension—a sense that our most fundamental theories describe *what* happens with extraordinary accuracy, but do not always explain *how* it happens in terms of tangible physical processes. The present volume, **The MEDIUM Disclosure: Constitutive Physics of the Absolute Medium**, is written in response to that tension.

This work advances a single unifying proposal: that the Universe is permeated by a real, ultra-stiff, non-viscous continuum—the **Absolute Medium (AM)**—and that all known physical phenomena emerge from its elastic, hydrodynamic, and nonlinear responses. The AM is not a revival of the historical ether. It is a new mechanical substrate derived from first-principles equations, supported by modern constraints, and capable of reproducing gravitational, quantum, and cosmological behavior using a unified continuum framework.

Across the chapters of this book, the AM is developed not as metaphor, but as **constitutive physics**: a medium with definable stiffness, density, strain, pressure laws, nonlinear elastic behavior, vortex flow fields, saturation limits, and field tensors. Wherever possible, concepts are introduced **equation-first**, allowing the reader to observe how structure, force, curvature, confinement, coupling, and wave propagation arise from mechanical laws alone.

Within this framework, gravity becomes a hydrodynamic event—a push from dissociative pressure and sink-vortex inflow rather than a geometric pull. Quantum behavior emerges from nonlinear displacement fields, saturation thresholds, and pressure curvature.

Nuclear forces arise from near-field r^{-4} elastic amplification and density-driven volumetric coupling. Even cosmological expansion and re-seeding of matter follow naturally from the Medium's baseline pressure and phase transitions.

The material in this book is not speculative in the casual sense. It is the systematic development of a **fully mechanical theory of everything observable**, expressed through constitutive equations, conservation laws, derivations, and scaling relations. Every section—from AM vortices to field tensors, from strong-force reconstruction to quantum tunneling, from black-hole sinks to galaxy-scale SDL halos—is intended to provide both conceptual clarity and mathematical rigor.

The Absolute Medium framework does not ask the reader to discard established physics. Rather, it invites them to reinterpret it—placing familiar phenomena within a deeper, physically grounded substrate. The AM equations recover classical mechanics, electromagnetism, quantum features, and relativistic limits, not by analogy, but by construction. They offer the possibility that the abstractions of modern theories are emergent shadows of a more fundamental mechanical layer that has long been assumed absent.

This book represents the second installment of *The MEDIUM Disclosure* series. It focuses specifically on **constitutive physics**: the internal structure, laws, limits, and behaviors of the Absolute Medium itself. It is the author's conviction that only by specifying this Medium in explicit mechanical detail—its non-linear elasticity, volumetric coupling, stiffness, density, pressure, flow, and saturation—can we construct a coherent and predictive picture of the universe that seamlessly unifies the microscopic and the cosmic.

Readers approaching this work will find that it rewards both mathematical curiosity and physical intuition. Some sections read like fluid mechanics; others like solid-state elasticity; still others like quantum field theory, general relativity, or plasma physics. But throughout, one message remains constant: the universe is not empty. It is built upon a structured, elastic, dynamic continuum whose laws govern everything we observe—from the spin of planets to the stability of nucleons, from the cosmic microwave background to the emergence of quantum interference.

This book is an invitation—to reconsider the foundations of physics, to explore a unifying medium beneath phenomena, and to participate in the unfolding development of a mechanical cosmology grounded in constitutive law rather than abstraction. It is offered with gratitude to all those who continue to question, to derive, to calculate, and above all, to imagine a universe that is intelligible not only mathematically, but physically.

Introduction: Why an Ultra-Dense Medium Appears Empty

The Absolute Medium (AM) framework proposes that spacetime is not a vacuum but an ultra-dense, ultra-stiff elastic continuum. Its background density ($\approx$10^17 kg/m^3) and stiffness far exceed those of known matter. Yet at human scales, the medium appears perfectly frictionless, exerting no detectable resistance. This apparent contradiction is the Observability Paradox.

The paradox is resolved by recognizing that the medium's interaction with matter depends strongly on scale and concentration. Three principles form the explanatory foundation:

1. Scale-Dependent Rigidity (SDRP)

2. Coupling Hierarchy Law (CHL)

3. Nucleon Density Constraint (NDC)

Scale-Dependent Rigidity (SDRP)

The AM's intrinsic stiffness is enormous, but its effective stiffness depends on the magnitude of displacement gradients. Large objects generate extremely small gradients, so the medium appears soft. At nucleon scales, gradients become large, and the medium exhibits its full rigidity. This explains why macroscopic matter experiences no drag, while nucleons interact strongly with the medium.

Coupling Hierarchy Law (CHL)

Matter's coupling to the AM is not uniform; it scales with local nucleon density and inverse system size. Dilute matter at human scales has negligible coupling strength (far below detectability), while nucleon-scale concentrations couple strongly. This hierarchy spans more than 30 orders of magnitude and explains the absence of ether-drift effects.

Nucleon Density Constraint (NDC)

Stable nucleons require that the medium's intrinsic density exceed their internal energy density. Modeling nucleons as localized AM deformations yields a required AM density on the order of 10^17 kg/m^3. Thus, the medium's density is not an arbitrary assumption—it is dictated by nucleon stability.

RESOLVING THE PARADOX

Together, the three principles show that:

- The AM must be ultra-dense to support nucleons.
- It must appear frictionless at human scales due to weak coupling.

• It must exert strong influence at nuclear scales through nonlinear elasticity.

Implications

This framework provides mechanical interpretations of nuclear forces, radioactive decay, and potential medium-based technological applications such as controlled decay manipulation.

The AM Observability Paradox is resolved once scale-dependent rigidity, concentration-dependent coupling, and nucleon stability constraints are recognized as natural consequences of nonlinear medium elasticity. What appears as empty vacuum at human scales is, in fact, a dense, structured continuum essential to the existence of matter itself.

Dissociative Pressure

1. Fundamental Equation of Dissociative Pressure

In the AM Model, Dissociative Pressure (P_D) is the universal compressive stress exerted by the Absolute Medium upon any region of reduced density (matter). The governing form is:

- P_D(r) = k * (Δρ)^2
- where Δρ is the density differential between the Absolute Medium and the matter-induced displacement, and k is the Medium's non-linear elastic constant (NLEP coefficient).

2. Near-Field Non-Linear Pressure Gradient (r^{-4} Law)

In the vicinity of a mass, the Medium's stiffness yields a pressure gradient described by:

- $P_D(r) \propto 1 / r^4$
- This arises from the squared strain response of the Absolute Medium as matter displaces the Energetic Net.

3. Mechanical Interpretation

Dissociative Pressure attempts to collapse matter's low-density lattice, driving sink flow and powering the AM vortex. It is the primary motivator of gravitational inflow in the Medium-centric framework.

4. Relation to Matter Stability

Matter remains stable when internal vibratory pressure balances Dissociative Pressure. If P_D exceeds nuclear structural capacity, dissociation begins (e.g., neutron decay, black hole formation).

Vortex

1. Fundamental Equation of the AM Vortex

In the AM Model, the vortex arises from the spiraling sink flow of the Absolute Medium toward a mass. The core equation governing this flow is:

- $v_\theta(r) = (\Gamma / 2\pi r)$
- where v_θ is tangential Medium velocity, r is radial distance, and Γ is circulation induced by Dissociative Pressure and volumetric coupling.

2. Vortex Driven by Sink Flow

The sink component of the vortex obeys:

- $v\_r(r) = - Q / (2\pi r)$
- where Q is the Medium inflow rate due to the mass-induced displacement kernel.

3. Combined Flow Field

The full AM vortex is the superposition of radial sink flow and angular twist:

- $v(r) = sqrt(v\_r(r)^2 + v\_\theta(r)^2)$

4. Physical Interpretation

These equations describe the mechanical spiraling of the Absolute Medium into a mass, producing rotation of planets and orbital dynamics via volumetric coupling.

Stiffness

1. Fundamental Definition of Medium Stiffness

In the AM Model, Stiffness (K_AM) expresses how strongly the Absolute Medium resists displacement. The core constitutive form is:

- $K\_AM = dP / d\varepsilon$
- where P is Medium pressure and ε is volumetric strain induced by matter. High K_AM reflects the ultra-rigid Energetic Net.

2. Non-Linear Elastic Response (NLEP Link)

Because the Medium follows the Non-Linear Elasticity Principle, stiffness scales with strain as:

- $P(\varepsilon) = k * \varepsilon^2$
- Thus: $K_{AM} = dP/d\varepsilon = 2k\varepsilon$
- This means the Medium stiffens as displacement increases, producing near-infinite resistance near matter cores.

3. Stiffness–Wave Speed Relation

Light is a transverse vibration of the Energetic Net. Its speed c is governed by the stiffness-to-inertia ratio:

- $c = \sqrt{K_{AM} / \rho_{AM}}$
- The extremely high K_{AM} and extremely low ρ_{AM} give rise to the cosmic speed limit.

4. Mechanical Interpretation in AM Model

Stiffness allows the Medium to transmit forces rapidly, maintain structural continuity, provide instantaneous gravitational coupling, and grip matter volumetrically (VCP).

Elasticity

1. Fundamental Definition of Medium Elasticity

Elasticity (E_{AM}) characterizes the Medium's ability to return to its equilibrium configuration after deformation. The constitutive definition is:

- $E_{AM} = d\sigma / d\varepsilon$
- where σ is elastic stress in the Absolute Medium and ε is strain produced by matter-induced displacement.

2. Non-Linear Elastic Response (NLEP)

In NLEP, the Medium responds quadratically to displacement, leading to:

- $\sigma(\varepsilon) = k * \varepsilon^2$
- Hence the effective elasticity becomes: $E_{AM}(\varepsilon) = d\sigma/d\varepsilon = 2k\varepsilon$
- Elasticity increases with strain, yielding the steep r^{-4} near-field gravitational behavior.

3. Restorative Dynamics and Snap-Back Behavior

Elasticity supplies the restoring force resisting matter displacement, producing the Medium "snap-back":

- $F_{elastic} \propto E_{AM} * \varepsilon$
- This maintains the Medium's structural continuity and drives gravitational restoration.

4. Elasticity and Wave Propagation

The Medium's elasticity determines propagation of electromagnetic disturbances:

- $c = \mathrm{sqrt}(K_{AM} / \rho_{AM})$
- Elasticity (paired with stiffness) enables rapid, low-loss propagation across cosmic scales.

L_fund

L_fund represents the grain scale of the Energetic Net — the smallest physically meaningful length within the AM continuum. Its existence establishes ultraviolet cutoff, zero-point oscillation frequency, renormalization boundaries, and the mechanical substrate of quantum effects.

1. Definition of the Fundamental Length

In the AM Model, L_fund is the characteristic lattice spacing of the Energetic Net. It defines the highest possible spatial resolution of Medium structure.

$L_{fund} = (\pi c) / \omega_{fund}$ (1)

2. Maximum Admissible Wavenumber and Frequency

Because L_fund is the smallest meaningful AM length, it sets an upper bound on wavenumber k and angular frequency ω:

$k_{max} = \pi / L_{fund}$ (2)

$\omega_{max} = c\, k_{max} = (\pi c) / L_{fund}$ (3)

3. Zero-Point Breathing Modes

Each AM grain oscillates at the natural frequency ω_{fund}, producing the baseline zero-point energy:

$E_0 = (1/2)\, \hbar\, \omega_{fund}$ (4)

4. Zero-Point Energy Density and Pressure

The grain volume is $V_{fund} = L_{fund}^3$. The zero-point energy density and corresponding AM baseline pressure are:

$$\rho_0 = E_0 / L_{fund}^3 \quad (5)$$

$$P_\Lambda = \xi \rho_0 \quad (6)$$

5. Ultraviolet Renormalization Cutoff

The spectral cutoff in the AM field theory eliminates ultraviolet divergences by enforcing the limit $k \leq k_{max}$. A typical renormalized quantity such as zero-point energy integral becomes finite:

$$E_0(\text{total}) = (\hbar c / 16\pi^2) (\pi / L_{fund})^4 \quad (7)$$

6. Influence on Nonlinear Elasticity (NLEP)

The Non-Linear Elasticity Principle implies local stress $\sigma = k \varepsilon^2$. At scales smaller than L_{fund}, strain gradients saturate. Thus NLEP modifies to prevent divergence:

$$\sigma_{max} = k \varepsilon_{max}^2 \quad (8)$$

7. Implications for Quantum Effects

Quantum tunneling, interference, vacuum drag, and saturation collapse receive mechanical grounding from L_{fund} through pressure curvature and grain-scale oscillations.

8. Summary

L_fund provides the physical cutoff of the AM, defines zero-point structure, supports renormalization, regulates nonlinear elasticity, and anchors the mechanical substrate of quantum-classical behavior.

M_crit

M_crit marks the transition beyond which translational speed no longer contributes to internal structural stability. Beyond this boundary, internal stability is saturated; velocity serves only as motion, not as a stabilizing factor. Governing physics transitions from internal mass-shielding and kinetic stabilization to external energy flows and Intrinsic Kinetic Resistance (IKR).

1. Formal Definition

In the AM Model, the Critical Mass Boundary M_crit is the minimal mass of a bound kernel for which internal elastic and volumetric coupling mechanisms reach saturation and can no longer be strengthened by increasing the body speed relative to the Medium. Beyond this point, speed provides no additional stiffening or shielding; stability depends on ambient AM pressure and external energy exchange.

Adopted postulate (AM archives):

$M\_crit \approx 97.6 \cdot m\_n$

where m_n is the nucleon mass (formal postulate recorded in AM model memory).

2. Mechanical Criterion at the Boundary

Let sigma_max denote the maximum elastic counter-stress the AM can provide at the kernel boundary (saturation at the grain scale), and let P_D be the local AM dissociative pressure generated by the kernel density deficit. The boundary is reached when incremental kinetic stiffening vanishes:

d(sigma_eff)/dv at M = M_crit equals 0

Equivalently, the near-field stabilization balance in pressure form becomes marginal:

P_D(M_crit, r) = sigma_max(r)

3. Constitutive Ingredients

Non-Linear Elasticity (NLEP):

sigma = k_NLEP · epsilon^2

Kernel displacement (nucleon-style scaling):

epsilon(r) = alpha · M / r^2

Near-field AM pressure (r^(-4) law):

P(r) = k_NLEP · (alpha M / r^2)^2 = k_NLEP · alpha^2 · M^2 / r^4

4. Scaling Derivation

Let r_* denote the effective kernel radius at which saturation first appears (typically close to the AM grain scale or kernel core scale). Setting the near-field pressure equal to the saturation stress yields:

k_NLEP · alpha^2 · M_crit^2 / r_*^4 = sigma_max

$$M\_crit = (r\_*^2 / \alpha) \cdot \sqrt{\sigma\_max / k\_NLEP}$$

This expression shows that M_crit is fixed by the micro-constitutive parameters (k_NLEP, sigma_max), the kernel geometry (via r_*), and the mapping constant alpha.

5. Velocity Decoupling Beyond M_crit

Below M_crit, finite velocity v can contribute to effective stiffening through dynamic coupling. Beyond M_crit, any additional v only increases motion and does not raise sigma_eff. The decoupling is expressed by a vanishing derivative with respect to v and the invariance of the stability functional:

$$S(M, v) = S(M) \text{ for } M \geq M\_crit \quad (9)$$

6. Intrinsic Kinetic Resistance (IKR) Regime

In the post-critical regime, transport and deformation are governed by Intrinsic Kinetic Resistance (IKR), a Medium-induced inertial response to changes in the kernel state. A simple constitutive form is:

$$F\_IKR = \Gamma(M) \cdot a + \Lambda(M) \cdot \dot{a} \quad (10)$$

where Gamma(M) and Lambda(M) are effective mass-like and damping-like Medium coefficients calibrated at M ≥ M_crit.

7. Consistency with Nucleon-Scale Parameters

Using the nucleon mass m_n as the reference kernel and adopting the formal archive value M_crit ≈ 97.6 m_n, the boundary sits at a mesoscopic aggregate composed of O(10^2) nucleons. This scale is large enough to engage strong volumetric coupling yet small enough that the r^(-4) near-field remains decisive.

8. Implications and Regime Map

8.1 Nuclear structure and radioactivity: Beyond M_crit, additional kinetic energy does not improve containment; decay is driven by mechanical failure of AM confinement rather than lack of speed.

8.2 Dense-matter applications: In neutron-star analogues, the relevant mass elements are much greater than M_crit; engineering relies on pressure management (AM-lens) rather than kinematic stiffening.

8.3 Propulsion and shielding: For aggregates at or above M_crit, velocity does not harden internal structure—design must prioritize AM pressure gradients and IKR tuning.

9. Experimental/Observational Windows

• Mesoscopic cluster tests (N ≈ 50–200 nucleons) to detect a plateau in velocity-dependent stiffening.

• High-g acceleration of near-critical aggregates to verify d(sigma_eff)/dv → 0.

• IKR extraction via pulsed-acceleration response to calibrate Gamma(M) and Lambda(M) across the M ≈ M_crit neighborhood.

10. Notation and Parameters

• m_n — nucleon mass; M_crit — critical mass boundary.

• k_NLEP — non-linear elastic coefficient of AM; sigma_max — saturation stress at grain scale.

• alpha — kernel displacement constant (epsilon = alpha M / r^2); r_* — effective kernel radius at saturation onset.

• P_D — dissociative pressure; sigma_eff — effective stabilizing stress; v — translational speed.

• Gamma(M), Lambda(M) — IKR coefficients; a — acceleration; a_dot — time derivative of a.

11. Physical Summary

M_crit identifies the onset of saturation in AM-mediated stabilization: beyond this mass threshold, speed can no longer harden internal structure. The boundary follows from near-field r^(-4) pressure, NLEP saturation, and finite sigma_max at the grain scale, yielding a scaling form M_crit = (r_*^2 / alpha) · sqrt(sigma_max / k_NLEP). The adopted AM archive value M_crit ≈ 97.6 m_n places the transition at a mesoscopic nucleon count, with implications for nuclear mechanics, dense-matter physics, propulsion concepts, and AM-pressure engineering.

The Gravitational Event

1. Defining the Gravitational Event Mechanically

In the AM Model, a Gravitational Event is the mechanical consequence of the Absolute Medium responding to a matter-induced displacement. The core relation is:

- G_Event = { P_D + K_AM + E_AM + Vortex_Flow }
- where P_D is Dissociative Pressure, K_AM is Medium Stiffness, E_AM is Elastic Restorative Force, and Vortex_Flow is the spiraling sink component.

2. Pressure Driver: Dissociative Pressure

The gravitational inflow is initiated by the squared-density pressure gradient:

- $P\_D(r) = k\,(\Delta\rho)^2$
- and near matter surfaces:
- $P\_D(r) \propto 1 / r^4$

3. Transmission: Stiffness and Elastic Response

The Medium transmits gravitational influence via its stiffness and non-linear elasticity:

- $K\_AM = dP/d\varepsilon$
- $E\_AM = d\sigma/d\varepsilon = 2k\varepsilon$
- These relations define the Medium's ability to instantaneously communicate pressure imbalance.

4. Motion Component: Sink and Vortex Flow

Matter experiences motion because it is entrained in the sink flow of the Medium:

- $v\_r(r) = - Q / (2\pi r)$
- and rotationally via the circulation term:
- $v\_\theta(r) = \Gamma / (2\pi r)$

5. Total Gravitational Acceleration

The effective acceleration is governed by Medium–matter coupling and flow:

- $a(r) = (P\_D / \rho\_fluff) + (v\_r\, dv\_r/dr) + (v\_\theta^2 / r)$

- This equation expresses gravity as a hydrodynamic push from the Absolute Medium.

6. Physical Summary

A Gravitational Event is the unified outcome of pressure imbalance, Medium stiffness, elastic restoration, and vortex entrainment. Objects fall because they are carried by Medium inflow, not pulled by matter.

Clustering

1. Fundamental Pressure-Shielding Equation

In the AM Model, Clustering arises when two masses reduce the Absolute Medium pressure between them. The governing equation for the Pressure Shadow is:

- $\Delta P\_shadow = P\_outside - P\_between$
- where P_outside is unperturbed Dissociative Pressure and P_between is the reduced pressure due to mutual shielding.

2. Dissociative Pressure Driver (Near-Field Form)

The collapse-driving pressure is given by the non-linear form:

- $P\_D(r) = k\,(\Delta\rho)\^2$
- with near-field gravitational amplification:
- $P\_D(r) \propto 1 / r\^4$
- Thus, when two masses approach, the r^{-4} terms overlap, reducing internal pressure more steeply.

3. Medium Flow Superposition: Sink Merging

Each mass generates a sink inflow given by:

- v_r(r) = - Q / (2πr)
- Clustering occurs when the superposed sink flows produce an attractive convergence:
- v_total = v_r1 + v_r2
- This creates a merged flow corridor drawing bodies together.

4. Force of Clustering (Medium Push)

The effective clustering force is a Medium-pressure push, not a pull:

- F_cluster = A * ΔP_shadow
- where A is the effective cross-sectional area over which the Medium applies the push.

5. Stability Condition for Clusters

Stable clustering occurs when the inward Medium push exceeds orbital dispersive effects:

- F_cluster ≥ m v^2 / r
- ensuring the bodies remain bound within the shared sink system.

6. Physical Summary

Clustering in the AM Model is the natural result of pressure-shadowing, sink-flow merging, and Medium-driven push. Matter aggregates because the Medium prefers minimizing distributed displacement zones, forming galaxies, clusters, and large-scale structure.

Mass-Shielding

1. Pressure-Shadow Formation (Primary Equation)

Mass-shielding arises when one mass blocks incoming Absolute Medium flow to another. The core relation is:

- ΔP_shadow = P_D(outside) - P_D(between)
- where P_D is the Dissociative Pressure exerted by the Medium.

2. Dissociative Pressure Contribution

The local dissociative pressure around each mass is:

- P_D(r) = k (Δρ)^2
- Near-field gravitational amplification:
- P_D(r) ∝ 1 / r^4
- Overlapping r^{-4} zones reduce the Medium pressure in the region between bodies.

3. Medium Flow Attenuation Through Fluffy Matter

Because matter is permeable (Volumetric Coupling), the Medium-flow attenuates according to:

- Q_out = Q_in * exp(- C_v * L)
- where C_v is the volumetric coupling coefficient and L is path length through the mass.

4. Shielding Factor Definition

The shielding effectiveness between two masses is quantified by:

- S = 1 - (Q_out / Q_in)
- High S indicates strong attenuation of the Medium flow, producing strong clustering or gravitational reinforcement.

5. Medium Push Resulting from Shielding

The Medium pushes masses together with force:

- F_MS = A * ΔP_shadow
- where A is the effective cross-section exposed to the differential pressure.

6. Scale-Dependent Mass-Shielding

- • Nanometric realm: shielding reduces vibrational noise → atomic stability.
  • Cosmic realm: shielding reduces pressure between planets/ stars → gravitational attraction.
  • High-density regimes: shielding saturates → neutron stars, black hole precursors.

7. Physical Summary

Mass-shielding in the AM Model is a mechanical reduction of Medium pressure between objects, produced by dissociative pressure gradients and attenuation of Medium flow. This creates an inward push, driving clustering and gravitational binding.

Mass-Shielding in Different Realms

1. Realm-Dependent Shielding Framework

Mass-shielding behaves differently depending on the scale of interaction. The general Medium attenuation law is:

- $Q_{out} = Q_{in} * \exp(- C_v * L)$
- where C_v is the Volumetric Coupling coefficient and L is the path length through matter.

2. Nanometric Realm: Vibrational Shielding

In the nanometric regime, shielding suppresses the Medium's grain-scale vibrational noise. Dominant equation:

- $S_{nano} = 1 - (Q_{out} / Q_{in})$
- Here, Q_{out} is reduced because attenuation is significant when particle size approaches L_{fund}.
- The resulting effect is vibratory silence within atomic cores, enabling matter stability.

3. Planetary Realm: Hydrodynamic Shielding

At planetary scales, shielding manifests as pressure reduction between large bodies:

- $\Delta P_{shadow} = P_D(\text{outside}) - P_D(\text{between})$
- with Dissociative Pressure:
- $P_D(r) \propto 1 / r^4$
- Overlapping r^{-4} fields create gravitational attraction via Medium push.

4. Stellar and Neutron Star Realm: Saturation Shielding

For ultra-dense matter, C_v becomes extremely high, creating near-total attenuation:

- $\lim_{\rho \to \rho_{max}} (Q_{out} / Q_{in}) \to 0$
- This yields maximal shielding, producing deep gravitational wells and stiff Medium curvature.

5. Black Hole Regime: Total Medium Exclusion

At the Dissociative Limit, shielding reaches unity:

- $S \to 1$
- Matter fully blocks Medium inflow; dissociation converts matter to raw Medium, creating a total sink.

6. Realm Comparison Summary

Realm	Shielding Behavior	Dominant Equation
Nanometric	Vibrational damping; stability of nucleons	$S = 1 - (Q_{out}/Q_{in})$
Planetary	Medium pressure shadow → gravitational push	$\Delta P_{shadow} = P_D(out) - P_D(in)$
Stellar / Neutron Star	Saturation shielding; extreme curvature	$\lim (Q_{out}/Q_{in}) \to 0$
Black Hole	Total shielding; dissociation	$S \to 1$

Volumetric Coupling Principle (VCP)

1. Definition of Volumetric Coupling

The Volumetric Coupling Principle quantifies how strongly the Absolute Medium (AM) couples to the internal volume of matter. It is governed by:

- $C_v = \kappa / L_{fund}$
- where κ is the geometric factor and L_{fund} is the Fundamental Length of the Energetic Net.

2. Medium–Matter Interaction Law

The Medium flow attenuation through matter follows:

- $Q_{out} = Q_{in} * \exp(- C_v * L)$
- where L is the characteristic path length through the matter volume.

3. Full Derivation of C_v from First Principles

Given the AM stiffness and density, the drag-like interaction for porous matter is derived from internal surface contact:

- $dQ = - (\kappa / L_{fund}) Q\, dx$
- Integrating: $\int(dQ/Q) = - (\kappa/L_{fund}) \int dx$
- Yielding: $Q_{out} = Q_{in} * \exp(- \kappa L / L_{fund})$

4. Coupling-Induced Force Density

The force density exerted by the Medium on matter is:

- $f = C_v * P_D$
- with P_D the dissociative pressure $P_D = k (\Delta\rho)^2$.

5. Volumetric Effective Cross Section

Matter's effective cross section for Medium coupling is:

- $A_{eff} = \int \rho_{local}\, dV$
- and total coupling becomes: $\Gamma = C_v * A_{eff}$

6. Gravitational Application

VCP defines gravitational entrainment through:

- $a(r) = (\Gamma / m) * (P_D / \rho_{AM})$
- linking Medium stiffness, density, and flow directly to observed acceleration.

VCP Evolution with Matter Density

1. Fundamental Density-Dependence of VCP

The efficiency of the Medium–matter handshake changes with local matter density (ρ_m). The core relation is:

- $C_v(\rho_m) = (\kappa / L_{fund}) * f(\rho_m)$
- where $f(\rho_m)$ is the density-dependent enhancement factor.

2. Low-Density Regime ($\rho_m << \rho_{crit}$)

For ordinary atomic matter, internal porosity is high and coupling scales approximately linearly:

- $f(\rho_m) \approx \rho_m / \rho_{ref}$

- yielding weak Medium grip and low gravitational response.

3. Intermediate Regime: Compression Begins

- As matter becomes denser, internal pore pathways shrink and coupling strengthens:
- $f(\rho_m) \propto \rho_m^n$ with $1 < n < 2$
- This accounts for strong gravity of terrestrial planets compared to gas clouds.

4. High-Density Regime ($\rho_m \to \rho_{sat}$)

- At extreme density, pathways close and coupling saturates:
- $\lim_{\rho_m \to \rho_{sat}} f(\rho_m) \to \infty$
- This produces:

- Welded Medium–matter interface
- High surface drag
- Ultra-strong local curvature

5. Dissociative Limit: Breakdown of Structure

- When matter density exceeds structural capacity, the Medium no longer flows through but crushes the lattice:
- $C_v \to \infty$
- This triggers dissociation and formation of a Pure Sink (Black Hole).

6. Combined Formulation

A simple interpolating model describing C_v evolution is:

- $C_v(\rho_m) = (\kappa / L_{fund}) * (1 + (\rho_m / \rho_{ref})^n)$
- with n controlling the steepness of the density transition.

7. Summary

As matter density increases, its volumetric engagement with the Absolute Medium strengthens from weak permeable flow to full saturation. This density-driven evolution of VCP explains planetary gravity, neutron-star rigidity, and black-hole dissociation in a unified mechanical framework.

The Breaking Point / Black Hole

1. Definition of the Breaking Point

The Breaking Point is reached when matter density becomes sufficient to overwhelm the Medium's elastic support. The criterion is:

- $P_D(r) \geq P_{EL,max}$
- where P_D is Dissociative Pressure and $P_{EL,max}$ is the maximum elastic counter-pressure of the Absolute Medium.

2. Dissociative Pressure at Critical Radius

Dissociative Pressure grows non-linearly as:

- $P_D(r) = k (\Delta\rho)^2$
- with near-field divergence:
- $P_D(r) \propto 1 / r^4$
- At sufficiently small r, the Medium's restoring ability collapses.

3. Medium Saturation Condition

The Absolute Medium reaches saturation when strain exceeds its structural limit:

- $\varepsilon \geq \varepsilon_{max}$
- where ε is Medium strain due to matter displacement.
- Using the NLEP relation:
- $\sigma(\varepsilon) = k \varepsilon^2$
- the corresponding saturation stress is:
- $\sigma_{max} = k \varepsilon_{max}^2$

4. Collapse Criterion for Black Hole Formation

A Black Hole forms when the total Medium stress is insufficient to support the displacement kernel. The condition is:

- $P_D(r_{crit}) > \sigma_{max}$
- This defines the critical radius r_{crit} at which matter collapses into a Pure Sink.

5. Pure Sink Condition

At the critical limit, Medium flow becomes singular:

- $v_r(r) \to \infty$
- and volumetric coupling diverges:
- $C_v \to \infty$
- Thus matter is fully dissociated into the raw Medium.

6. Event Horizon as Flow-Speed Boundary

The Event Horizon arises where inward Medium flow speed equals the Medium wave speed:

- $v_r(r_H) = c$
- with:
- $c = \sqrt{K_{AM} / \rho_{AM}}$
- Inside r_H, no ripple (light) can propagate outward.

7. Physical Summary

A Black Hole is the mechanical consequence of the Medium losing its ability to elastically sustain the displacement imposed by matter. The Breaking Point marks the transition from structured matter to Pure Sink, governed by the non-linear divergence of Dissociative Pressure and saturation of Medium elasticity.

The Medium Exhaust / Jets

1. Conservation of Medium Flow (Sink–Source Balance)

A Black Hole acts as a Pure Sink of the Absolute Medium. Conservation of Medium flow requires an equal Source (Exhaust). The governing continuity equation is:

- $Q_{in} = Q_{out}$
- where Q_{in} is the total Medium inflow through the sink boundary and Q_{out} is the total ejected flow through the polar exhaust.

2. Compression of Medium at the Core

The inflowing Medium is compressed near the collapse radius r_{crit}. Compression follows:

- $P_{core} \approx k (\Delta\rho)^2$

- with $\Delta\rho$ representing the density deficit created by matter dissociation.

3. Polar Jet Formation (Nozzle Dynamics)

The Medium is expelled preferentially through the lowest-resistance axis (the poles). Jet velocity arises from pressure release:

- $v_{jet} = \sqrt{2 (P_{core} - P_{env}) / \rho_{AM}}$
- where P_{env} is the ambient Medium pressure outside the jet.

4. Angular Momentum Transfer (Toroidal Recirculation)

The inflowing Medium carries angular momentum Γ from the galactic vortex. The toroidal recirculation condition is:

- $\Gamma_{in} = \Gamma_{out}$
- ensuring the jet maintains rotational coherence.

5. Jet Collimation from Medium Stiffness

The extreme stiffness K_{AM} of the Medium confines the jet into a narrow column:

- $\theta_{jet} \propto 1 / \sqrt{K_{AM}}$
- where θ_{jet} is the opening angle. Higher stiffness produces stronger collimation.

6. Energy Output of the Medium Exhaust

Jet power is given by:

- $L_{jet} = (1/2)\, \rho_{AM}\, v_{jet}^3\, A_{jet}$

- with A_jet the jet cross-sectional area. This power corresponds to the energy released as dissociated matter returns to the Medium.

7. Physical Summary

Medium Exhaust (Jets) arises from pressure release at the Pure Sink boundary. The Absolute Medium, compressed at the Breaking Point, escapes through polar axes, conserving flow and angular momentum. Jets represent the re-expansion phase of the Medium, enabling the universal recycling of energy and maintaining the global sink–source balance.

Ingress Through Equatorial Plane / Egress Through Poles

1. Mass-Induced Flow Geometry (Baseline Equation)

In the AM Model, a rotating mass establishes an anisotropic pressure gradient. The Medium flow lines obey:

- $\nabla P\_D = - \rho\_AM * a\_flow$
- where P_D is Dissociative Pressure and a_flow is the induced Medium acceleration field.

2. Equatorial Ingress Condition

The inflow is maximized along the equatorial plane where rotational shear is highest. The radial inflow velocity is:

- $v\_in(r, \theta = \pi/2) = - Q / (2\pi r)$
- with Q the volumetric sink rate. Rotation contributes a twist

term:

- $v_\theta = \Gamma / (2\pi r)$
- causing spiraling equatorial ingress.

3. Polar Egress Condition

Medium compressed at the core escapes via the lowest-resistance axis (the poles). Jet velocity is given by:

- $v_{out} = \sqrt{2 (P_{core} - P_{env}) / \rho_{AM}}$
- where P_{core} is core Medium pressure and P_{env} is ambient pressure.

4. Continuity of Flow (Sink–Source Balance)

The total inflow over the equatorial belt equals the total ejected flow through the poles:

- $\oint_{equator} \rho_{AM} v_{in} dA = 2 \oint_{poles} \rho_{AM} v_{out} dA$
- ensuring global mass–flow conservation.

5. Magnetic Field Generation (Toroidal–Poloidal Coupling)

The circulation of the Medium forms a dynamo-like structure. The governing relation is:

- $B \propto \nabla \times v_{total}$
- where $v_{total} = v_{in} + v_\theta + v_{out}$.

6. Relation to Planetary Rotation

Ingress torque on matter follows:

- $\tau = \int \rho\_AM\ v\_in\ v\_\theta\ r\ dA$
- This torque sustains planetary spin through continuous Medium coupling.

7. Physical Summary

Equatorial ingress and polar egress arise from rotation-induced anisotropy in the Medium flow. The equator becomes a high-shear intake zone, while the poles become high-velocity exhaust nozzles. This geometry underlies gravity, magnetism, spin, and jet formation in the AM framework.

Torque & Planetary Spin

1. Medium-Induced Torque (Fundamental Relation)

Planetary spin arises from hydrodynamic torque exerted by the Absolute Medium flowing through the equatorial region. The basic torque expression is:

- $\tau = \int (\rho\_AM * v\_in * v\_\theta * r)\ dA$
- where v_in is equatorial ingress velocity, v_θ is angular (twist) flow, r is lever arm, and ρ_AM is Medium density.

2. Radial Inflow Component

The inward Medium velocity at the equator follows sink-flow dynamics:

- $v\_in(r) = - Q / (2\pi\ r)$
- with Q the volumetric intake rate determined by planetary mass and displacement.

3. Circulation-Induced Twist

The Medium carries angular momentum from the larger stellar vortex. The twist velocity is:

- $v\_\theta(r) = \Gamma / (2\pi\ r)$
- where Γ is circulation inherited from the stellar AM vortex.

4. Full Torque Integration

Substituting v_in and v_θ into the torque integral gives:

- $\tau = \int (\rho\_AM * (Q/(2\pi r)) * (\Gamma/(2\pi r)) * r)\ dA$
- Simplifying:
- $\tau \propto (\rho\_AM * Q * \Gamma) \int dA / r\text{^}1$
- Torque increases as 1/r near the equator, where flow is most intense.

5. Steady-State Rotation Rate

The planet reaches a stable spin when Medium-applied torque balances internal viscous resistance:

- $\tau\_in = \tau\_resist$
- with τ_resist determined by planetary density, internal friction, and structural rigidity.

6. Angular Acceleration from Medium Coupling

The angular acceleration α follows:

- $\alpha = \tau / I$
- where I is the planet's moment of inertia. High Q and Γ

produce rapid spin-up for gas giants.

7. Summary

Planetary spin arises from the Absolute Medium's spiraling inflow at the equator. The combination of radial intake and angular circulation produces continuous torque, aligning with the observed universal rotation of planets, stars, and accretion structures.

The Universal Loom

1. Sink–Source Circulation Equation (Foundational Loom Relation)

The Universal Loom describes the continuous circulation of the Absolute Medium (AM) through celestial bodies. The baseline conservation law is:

- $Q_{in} = Q_{out}$
- where Q_{in} is the equatorial ingress inflow and Q_{out} is the polar egress outflow.

2. Equatorial Ingress (Warp Flow) Equation

Medium intake along the equatorial belt follows sink-flow mechanics:

- $v_{in}(r) = - Q / (2\pi r)$
- and carries angular twist:
- $v_\theta(r) = \Gamma / (2\pi r)$
- This establishes the "warp threads" of the Loom.

3. Core Compression and Rebound Equation

The Medium compresses as it converges toward the core. Compression obeys:

- $P_{core} = k (\Delta\rho)^2$
- where $\Delta\rho$ is the density deficit introduced by matter.
- Elastic rebound from the core is governed by:
- $P_{EL} = 2k \varepsilon$
- from the non-linear elasticity principle (NLEP).

4. Polar Egress (Weft Flow) Equation

Medium exits through the poles via pressure-driven jets:

- $v_{out} = \sqrt{2 (P_{core} - P_{env}) / \rho_{AM}}$
- This forms the "weft threads" of the Loom.

5. Recirculatory Field Tensor of the Loom

The Universal Loom's full circulation can be written as a flow tensor:

- $L_{ij} = \partial_i v_j - \partial_j v_i$
- where v_j includes v_{in}, v_θ, and v_{out} components.

6. Magnetic Field Emergence (Toroidal–Poloidal Coupling)

Magnetic field generation arises from the curl of the Loom flow:

- $B \propto \nabla \times v_{total}$
- where $v_{total} = v_{in} + v_\theta + v_{out}$.

7. Matter Weaving Condition

Matter stability emerges from the balance of sink-driven compression and elastic rebound:

- P_D = P_EL
- where P_D = k (Δρ)^2 is Dissociative Pressure.

8. Physical Summary

The Universal Loom is the global circulatory engine of the Absolute Medium. Equatorial ingress draws Medium inward, core compression energizes it, and polar egress ejects it back outward. This cyclical motion underlies gravity, magnetism, spin, and matter stability across cosmic scales.

Static–Dynamic Duality Principle (SDDP)

1. Dual-State Definition of the Absolute Medium

The Absolute Medium exhibits simultaneous static and dynamic behavior. This duality is formalized as:

- AM_state = { Static_Net , Dynamic_Flow }
- where the Static_Net governs structural elasticity and the Dynamic_Flow governs sink/vortex motion.

2. Static Component: Elastic–Stiff Framework

The static part of the Medium behaves as an ultra-stiff elastic continuum. Its constitutive relations are:

- $\sigma = k\, \varepsilon^2$
- $K_{AM} = dP/d\varepsilon$
- These relations define a rigid Energetic Net that sets the universal wave speed:
- $c = \sqrt{K_{AM} / \rho_{AM}}$

3. Dynamic Component: Sink–Vortex Flow

The dynamic aspect describes hydrodynamic motion of Medium toward mass concentrations. The radial and tangential components are:

- $v_r(r) = - Q / (2\pi r)$
- $v_\theta(r) = \Gamma / (2\pi r)$
- These generate gravitational inflow and rotational entrainment.

4. Coupling Between Static and Dynamic Modes

Matter interacts with both states simultaneously via Volumetric Coupling. The dual coupling law is:

- $F_{total} = C_v (P_D + \rho_{AM} v^2)$
- where $P_D = k (\Delta\rho)^2$ defines static compression and $\rho_{AM} v^2$ defines dynamic transport.

5. Dual Transmission Equation (Instantaneous Reach + Flow Motion)

Static stiffness transmits pressure imbalances nearly instantly, while dynamic flow carries matter. The combined propagation law is:

- $a(r) = -\nabla P\_D / \rho\_AM + (v \cdot \nabla) v$
- expressing simultaneous elastic restoring force and hydrodynamic acceleration.

6. Gravitational Event as a Static–Dynamic Superposition

A gravitational event arises from overlapping static and dynamic fields:

- $G\_Event = \{ P\_D, K\_AM, E\_AM, v\_r, v\_\theta \}$
- Gravity emerges from the Medium pushing inward (static) while flowing inward (dynamic).

7. Physical Summary

The Static–Dynamic Duality Principle defines the Absolute Medium as a dual-mode continuum. Its static aspect provides structural rigidity and wave propagation, while its dynamic aspect produces gravitational inflow and rotational dynamics. Together they form a single, unified mechanical substrate underlying all physical interactions in the AM Model.

Structural Density Law (SDL: M^2/R^5)

1. Fundamental Statement of SDL

The Structural Density Law states that the density of the Absolute Medium at the surface of a mass M with radius R is:

- $\rho\_AM \propto M^2 / R^5$
- This reflects how mass-induced displacement compresses the Medium according to geometric and non-linear elastic factors.

2. Starting Point: Medium Displacement Kernel

Each mass M produces a displacement field ε(r) that decays with radius:

- $\varepsilon(r) \propto M / r^2$
- derived from volume deficit scaling with surface flux across spherical shells.

3. Applying Non-Linear Elasticity (NLEP)

The Medium reacts according to the non-linear elastic stress relation:

- $\sigma = k\,\varepsilon^2$
- Thus the induced pressure at radius r is:
- $P(r) \propto (M^2 / r^4)$
- This r^{-4} scaling matches the near-field gravitational pressure gradient.

4. Converting Pressure to Density via Stiffness

- Using the Medium stiffness relation K_AM = dP/dε, density increase Δρ_AM is proportional to induced pressure:
- $\Delta\rho\_AM \propto P / R$
- Substituting $P(R) \propto M^2 / R^4$ gives:
- $\Delta\rho\_AM \propto (M^2 / R^5)$

5. Final SDL Form

- Combining geometric and elastic constants into calibration constant C_SDL yields:
- $\rho\_AM(R) = C\_SDL * (M^2 / R^5)$
- This matches planetary, stellar, and galactic-scale observations.

6. Implication for Gravity and Flow

- The gravity field strength is linked to ρ_AM via:
- g ∝ ρ_AM * R
- Thus:
- $g \propto (M^2 / R^4)$
- explaining deep gravity wells of compact objects.

7. Physical Summary

SDL arises from combining geometric displacement ($1/r^2$), non-linear elasticity (ε^2 dependence), and stiffness mapping (P→ρ). The result is a universal scaling law governing gravitational structure and Medium density across all cosmic regimes.

Matter–Medium Dual View

1. Dual Ontology of Matter and the Medium

In the AM Model, matter is treated as a displacement kernel embedded within a continuous Absolute Medium. The dual relationship is expressed as:

- Matter_state = { ρ_fluff , ε_kernel }
- Medium_state = { ρ_AM , σ_AM(ε) , v_flow }
- where ρ_fluff is low-density matter structure, ε_kernel its displacement, and ρ_AM, σ_AM, v_flow describe the Medium.

2. Matter as a Displacement Kernel

A matter particle produces a displacement ε of the Energetic Net. The displacement field is:

- $\varepsilon(r) \propto M / r^2$
- This displacement defines the "particle" in the Medium-centric interpretation.

3. Medium Wave Generated by Matter Motion

Motion of the displacement kernel forces elastic deformation of the Medium. The Medium responds with a vibrational wave:

- $\psi(r,t) = A \exp[i(k \cdot r - \omega t)]$
- The wave ψ represents elastic oscillations of the stiff Medium, not a probabilistic abstraction.

4. Kernel–Wave Coupling Equation

The particle and its wave are inseparable because the Medium enforces the coupling:

- $F = C\_v (P\_D + \rho\_AM\ v^2)$
- where $P\_D = k\ (\Delta\rho)^2$ is Dissociative Pressure. The particle's motion necessarily produces a wave, and the wave guides the particle.

5. Dual Propagation (Matter Drift + Wave Guidance)

The displacement kernel follows the hydrodynamic flow of the Medium:

- v_kernel = v_flow
- while the wave propagates according to:
- $c = \mathrm{sqrt}(K\_AM / \rho\_AM)$
- The kernel is slow, the wave is fast; the wave alters the pressure

landscape the kernel follows.

6. Interference Mechanism via Medium Pressure Fields

When passing through multiple slits, the Medium ripple obeys the superposition principle:

- $\psi\_total = \psi\_1 + \psi\_2$
- The kernel trajectory follows the gradient:
- $a = -\nabla P\_D / \rho\_fluff$
- Thus, interference is a Medium phenomenon acting on a real displacement kernel.

7. Full Matter–Medium Dual View Representation

The dual nature of matter is encoded as:

- $\{Particle\} = \varepsilon\_kernel$
- $\{Wave\} = \psi(AM)$
- and the Medium enforces:
- Particle $\leftrightarrow$ Wave (inseparable)

8. Summary

In the AM Model, matter is not independent of space; it is a localized distortion in the Absolute Medium. Motion of this distortion creates waves, and these waves shape the subsequent motion. This provides a mechanical, Medium-based explanation of the traditional wave–particle duality.

Two Realms Transition (Nanometric ↔ Cosmic)

1. Realm Definition via Medium State Function

The AM Model defines two interaction regimes—Nanometric (Energetic Net) and Cosmic (Hydrodynamic Flow). The governing state function is:

- AM_state(r) = f(L_fund , r)
- where L_fund is the Fundamental Length and r is interaction scale.

2. Nanometric Realm ($r \approx L\_fund$)

When interaction radius approaches the Net grain, the Medium behaves as a stiff elastic lattice. The displacement response is:

- $\varepsilon(r) \propto M / r^2$
- Elastic stress in this regime is:
- $\sigma = k\,\varepsilon^2$
- The force experienced is dominated by non-linear elasticity and yields near–strong-force confinement.

3. Cosmic Realm ($r \gg L\_fund$)

At macroscopic separations, the Medium transitions to a hydrodynamic continuum. Flow is governed by:

- $v\_r(r) = - Q / (2\pi r)$
- $v\_\theta(r) = \Gamma / (2\pi r)$
- producing gravitational inflow and vortex motion.

4. Transition Function Across Realms

The efficiency of Medium coupling across scales is defined by:

- C_v(r) = (κ / L_fund) * g(r/L_fund)
- where g(x) → 1 for x ≫ 1 (cosmic) and g(x) → x^n for x ≈ 1 (nanometric).

5. Resulting Force Unification

The general AM force law across realms is:

- F(r) = C_v(r) * P_D(r)
- with P_D(r) = k $(\Delta\rho)^2$ and P_D ∝ $1/r^4$ in near-field.
- This yields weak gravity at cosmic scales and strong-force-like confinement at nanoscales.

6. Realm Fusion at Extreme Density

In neutron-star and pre-collapse states, realm distinction vanishes as:

- lim_{ρ_m→ρ_sat} C_v → ∞
- This produces saturation coupling and welded Medium–matter interfaces.

7. Physical Summary

The Two Realms Transition expresses how the same Absolute Medium manifests as an elastic Net at nanoscales and a flowing continuum at cosmic scales. A unified mechanical law emerges from scale-dependent coupling, explaining strong nuclear effects, conventional gravity, and extreme astrophysical behavior in one framework.

Non-Linear Elasticity Principle (NLEP): r^{-4} Pressure Law

1. Fundamental Constitutive Postulate of NLEP

The Absolute Medium (AM) responds to displacement ε with non-linear elastic stress σ given by:

- $\sigma(\varepsilon) = k \cdot \varepsilon^2$
- This is the defining NLEP relation, expressing second-order elasticity of the Energetic Net.

2. Pressure as Elastic Response to Matter-Induced Displacement

Matter produces displacement of the Medium according to:

- $\varepsilon(r) \propto M / r^2$
- Substitution into $\sigma(\varepsilon)$ yields local AM pressure:
- $P(r) = k \cdot (M^2 / r^4)$
- This directly produces the inverse-fourth-power law.

3. Full r^{-4} Pressure Law Derivation

Starting from $\varepsilon(r) \propto M/r^2$, we write explicitly:

- $\varepsilon(r) = \alpha M / r^2$
- Then:
- $P(r) = \sigma(\varepsilon) = k\,(\alpha M/r^2)^2 = k\alpha^2 M^2 / r^4$
- Thus:
- $P(r) \propto M^2 / r^4$

4. Non-Linear Elastic Pressure Gradient and Force

The Medium push on matter is proportional to the gradient of P(r):

- $F(r) = - dP/dr = 4k\alpha^2 M^2 / r^5$
- This gives the inverse-fifth-power force density, matching SDL scaling.

5. Linking NLEP to the M^2/R^5 Structural Density Law (SDL)

Evaluating P(r) at r = R yields the AM density at a body's surface:

- $\rho\_AM(R) \propto P(R)/R \propto (M^2/R^4)/R = M^2/R^5$
- Thus NLEP mathematically generates the SDL.

6. Saturation Condition (Neutron Stars & Breaking Point)

Saturation occurs when $\varepsilon \to \varepsilon\_max$ such that:

- $\sigma\_max = k\, \varepsilon\_max^2$
- If P(r) exceeds σ_max, dissociation ensues. Therefore:
- $M^2 / r^4 \geq \sigma\_max \Rightarrow r \leq r\_crit$
- This defines the Breaking Point boundary and black-hole formation threshold.

7. Summary

NLEP maps microscopic displacement to macroscopic AM pressure via a quadratic constitutive law. The r^{-4} profile and its gradient create gravity-like forces, define the Structural Density Law, and determine neutron-star and black-hole thresholds. NLEP is therefore the mathematical engine of the AM Model's entire gravitational spectrum.

Unified AM Field Equation (P_D + NLEP + VCP + SDL)

1. Constituent Field Components

The unified AM Field Equation combines four governing principles of the Absolute Medium:

- • Dissociative Pressure: P_D
- • Non-Linear Elasticity (NLEP): $\sigma = k\,\varepsilon^2$
- • Volumetric Coupling (VCP): C_v
- • Structural Density Law (SDL): $\rho\_AM \propto M^2 / R^5$

2. Dissociative Pressure Field

Dissociative Pressure arises from density contrast $\Delta\rho$ between matter and Medium:

- $P\_D = k\,(\Delta\rho)^2$
- Substituting $\Delta\rho \propto M/r^2$ yields:
- $P\_D(r) \propto M^2 / r^4$

3. NLEP Contribution to the Field

From $\sigma(\varepsilon) = k\,\varepsilon^2$ and $\varepsilon(r) = \alpha\,M / r^2$, the elastic field becomes:

- $\sigma(r) = k\,\alpha^2\,M^2 / r^4$
- NLEP ensures steep pressure gradients near matter.

4. Volumetric Coupling (VCP)

VCP regulates how strongly matter couples to the AM field:

- $C\_v = \kappa / L\_fund$
- and modifies effective force transmission as:
- $F\_eff = C\_v \cdot P\_D$

5. SDL Contribution

SDL links the AM density at the surface of a body to its mass and radius:

- $\rho\_AM(R) = C\_SDL\ (M^2 / R^5)$
- This calibrates gravitational field magnitude.

6. Unified Field Equation (Master Form)

Integrating all contributions yields the AM Field Equation:

- $F\_AM(r) = C\_v \cdot P\_D(r) + \sigma(r) + \rho\_AM(r)\ v^2$
- Substituting each term gives:
- $F\_AM(r) = (\kappa/L\_fund)(k\ M^2/r^4) + (k\ \alpha^2\ M^2/r^4) + (C\_SDL\ M^2/r^5)\ v^2$
- Collecting M^2/r^4 and M^2/r^5 terms yields the compact unified representation:
- $F\_AM(r) \propto (M^2 / r^4)(A\_1 + A\_2/r)$
- where A_1 and A_2 are calibration constants derived from κ, L_fund, α, and C_SDL.

7. Summary

The unified field equation shows that all gravitational, elastic, and flow behaviors arise from a single mechanical law driven by Medium displacement and its non-linear elasticity. The M^2/r^4 base term represents core AM pressure, while M^2/r^5 determines structural density and large-scale gravitational strength.

The Nature of the Medium & Cosmic Evolution

1. Absolute Medium: Definition via Constitutive Invariance

The Absolute Medium (AM) is defined as an invariant, non-relativistic reference continuum. Its defining property is a fixed elastic–stiff state:

- c = sqrt(K_AM / ρ_AM)
- where c is light speed, K_AM is bulk stiffness, and ρ_AM is baseline Medium density.

2. Primordial Medium: Pre-Matter Background State

The Medium predates matter and forms the substrate from which matter arises. Its baseline pressure is:

- P_Λ = constant
- representing the irreducible equilibrium pressure of the AM.

3. Anti-Matter-Like Behavior: Dissociative Response to Matter

The Medium reacts to matter as to a defect or void. The response force is governed by Dissociative Pressure:

- P_D = k $(\Delta\rho)^2$

- which pushes inward on matter, attempting to restore Medium continuity.

4. Expansion via Phase Transition of the Medium

Cosmic expansion arises from the transition of compressed Medium into low-density space. Expansion rate relates to energy density drop:

- $\dot{R}/R \propto$ sqrt(P_Λ / ρ_AM)
- describing a pressure-driven expansion.

5. Re-Seeding: Frothing of the Medium into Matter

Matter nucleates from AM via pressure-drop-induced phase change. Local nucleation condition:

- ΔP = P_core - P_void > P_thresh
- leading to crystallization into displacement kernels (nucleons).
- Matter birth rate correlates with density gradients:
- Γ_seed ∝ dρ_AM/dr

6. Unified Cycle Equation

The universal cycle (Sink → Dissociation → Exhaust → Re-Seeding) obeys:

- Q_in (Sink) = Q_out (Source)
- ensuring global Medium conservation.

7. Summary

The AM is Absolute (fixed stiff substrate), Primordial (origin of all matter), and anti-matter-like (dissolves matter). Cosmic expansion and re-seeding arise from pressure transitions within the Medium. Equation-first analysis shows the Universe as a closed-cycle pressure-driven system.

Re-Seeding as Nucleon Crystallization

1. Phase-Transition Trigger: Pressure Differential Condition

Re-Seeding begins when a high-pressure Medium domain decompresses into a low-density cosmic void. The nucleation trigger is:

- $\Delta P = P\_core - P\_void \geq P\_thresh$
- where P_core is jet/exhaust pressure, P_void is ambient void pressure, and P_thresh is the crystallization threshold.

2. Decompression and Grain Emergence

As the Medium decompresses, local density drops as:

- $\rho\_AM(r) \rightarrow \rho\_void$
- The decompression rate follows:
- $d\rho\_AM/dt \propto -\nabla P$

3. Nucleon Crystallization Kernel

A nucleon forms as a stable displacement kernel when:

- $\varepsilon\_kernel = \alpha / r^2$
- and AM elastic stress satisfies:

- $\sigma = k\, \varepsilon\_kernel^2$
- Crystallization occurs when σ balances Dissociative Pressure:
- $\sigma = P\_D = k\, (\Delta\rho)^2$

4. Stability Condition (Net–Kernel Lock-In)

A nucleon becomes stable when the Medium's elastic restoring force equals internal oscillation pressure:

- $P\_EL = P\_internal$
- Using NLEP:
- $P\_EL = 2k\, \varepsilon$

5. Density Gradient Driving Nucleation

Nucleation rate depends on local AM density gradient:

- $\Gamma\_seed \propto d\rho\_AM/dr$
- Steeper gradients increase crystallization likelihood.

6. Kernel Aggregation into Atoms

Nearby nucleation sites merge when their displacement fields overlap:

- $\varepsilon\_total = \Sigma\, \varepsilon\_i$
- Binding occurs when:
- $\sigma\_total \geq P\_D(local)$

7. Summary

Re-Seeding is the crystallization of nucleons from decompressed Absolute Medium. Pressure drops create displacement kernels which stabilize under NLEP and Medium stiffness. These kernels aggregate into atoms, seeding new cosmic structure.

Re-Seeding → Atomic Lattice Formation

1. Medium Decompression and Kernel Nucleation

Re-Seeding begins with decompression of high-pressure Absolute Medium (AM) into low-pressure voids. The nucleation trigger is:

- $\Delta P = P\_core - P\_void \geq P\_thresh$
- This induces crystallization of displacement kernels from decompressed Medium.

2. Formation of a Nucleon Kernel

A nucleon forms as a stable displacement kernel when its internal strain profile is:

- $\varepsilon\_kernel(r) = \alpha / r^2$
- and Medium elastic stress satisfies:
- $\sigma = k\ \varepsilon\_kernel^2$
- Crystallization is achieved when $\sigma = P\_D = k(\Delta\rho)^2$.

3. Overlap and Binding of Multiple Kernels

Spatially adjacent kernels interact through overlapping displacement fields:

- $\varepsilon\_total(r) = \Sigma\ \varepsilon\_i(r)$
- Binding condition: $\sigma\_total \geq P\_D(local)$
- This produces the basic "pairwise bonding" mechanism between nucleons.

4. Proto-Lattice Formation Driven by AM Pressure Minimization

The Medium seeks to minimize total elastic energy. The arrangement that minimizes:

- $E\_total = \int k\, \varepsilon\_total^2\, dV$
- is a geometric lattice of evenly spaced kernels.

5. Stability Criterion of Emerging Atomic Lattice

A stable atomic lattice forms when elastic restoring forces balance dissociative pressure:

- $P\_EL = P\_D$
- Using NLEP: $P\_EL = 2k\, \varepsilon$
- Thus the stable lattice spacing d satisfies:
- $2k\, (\alpha/d^2) = k\, (\Delta\rho)^2$
- yielding: $d \propto 1 / \Delta\rho$

6. Atomic Shells from Hierarchical Kernel Interaction

Secondary displacement fields generate radial shell quantization:

- $\varepsilon\_n(r) = \alpha\_n / r^2$
- Shell stability occurs when adjacent modes satisfy:
- $\sigma\_n = \sigma\_\{n+1\}$

7. Unified Mapping from Re-Seeding → Atomic Lattice

Combining kernel formation, overlap, and lattice spacing gives:

- Lattice_formation = f(ΔP , ε_kernel , Σ ε_i , P_EL , P_D)
- Expanded:
- Lattice_formation ∝ (P_core − P_void) · (α/r^2) · (Σ α_i/r_i^2) · exp(−C_v L)
- This describes the complete mechanical sequence from Medium frothing to atomic lattice solidification.

8. Summary

Re-Seeding produces nucleons from decompressing Medium via NLEP-governed crystallization. Overlapping displacement kernels bind, forming proto-lattices that stabilize when elastic forces balance dissociative pressure. Atomic shells emerge from hierarchical displacement modes, completing the formation of matter.

Strong-Force Emergence from NLEP + VCP

1. Foundations: NLEP + VCP Coupling

The strong-force analogue in the AM Model emerges from the combined action of:

- • Non-Linear Elasticity Principle (NLEP): $\sigma = k\,\varepsilon^2$
- • Volumetric Coupling Principle (VCP): C_v = κ / L_fund · f(ρ_m)
- where ε is displacement strain, k is the AM elastic coefficient, and f(ρ_m) captures density-dependent coupling.

2. Near-Field Elastic Displacement

Matter kernels displace the Medium according to:

- $\varepsilon(r) = \alpha M / r^2$

Substitution into NLEP gives near-field pressure:

- $P\_NLEP(r) = k (\alpha M/r^2)^2 = k \alpha^2 M^2 / r^4$
- This r^{-4} pressure profile is the key to strong short-range attraction.

3. Density-Driven Volumetric Coupling Enhancement

As particle separation approaches nucleon scale, effective matter density rises and:

- $C\_v \to \infty$ as $\rho\_m \to \rho\_sat$
- This enhances force transmission dramatically.

4. Effective Strong-Force Law

The effective interaction between two nucleon kernels is:

- $F\_eff(r) = C\_v \cdot P\_NLEP(r)$
- Substituting C_v and P_NLEP gives:
- $F\_eff(r) = (\kappa/L\_fund \cdot f(\rho\_m)) \cdot (k \alpha^2 M^2 / r^4)$
- Thus:
- $F\_eff(r) \propto f(\rho\_m) \cdot (M^2 / r^4)$
- For dense matter f(ρ_m) grows steeply, giving strong-force-like magnitudes.

5. Saturation and Short-Range Cutoff

Saturation occurs when $\varepsilon \rightarrow \varepsilon\_max$ and Medium elasticity reaches its limit:

- $\sigma\_max = k\ \varepsilon\_max^2$
- When P_NLEP(r) ≥ σ_max, the force plateaus, producing the characteristic strong-force saturation.

6. Repulsive Core at Ultra-Short Range

At extremely small separation r, displacement gradients reverse sign, giving:

- $d\varepsilon/dr \rightarrow +\infty$

which yields an effective repulsive core:

- $F\_rep(r) \propto + dP\_NLEP/dr$

7. Full Strong-Force Expression (AM Model)

Combining attractive and repulsive terms gives:

- $F\_strong(r) = C\_v\ k\ \alpha^2\ M^2 / r^4 - d/dr\ (k\ \alpha^2\ M^2 / r^4)$
- Expanded:
- $F\_strong(r) = (\kappa\ k\ \alpha^2\ M^2 / L\_fund)\ f(\rho\_m)/r^4 - (4k\ \alpha^2\ M^2 / r^5)$
- This produces: short-range repulsion, mid-range strong attraction, long-range suppression.

8. Physical Summary

The strong force in the AM Model arises from the combined effects of the Non-Linear Elasticity of the Absolute Medium and density-driven Volumetric Coupling. The r^{-4} pressure field provides intense attraction, VCP amplifies it at nucleon densities, and NLEP saturation introduces the repulsive core. This reproduces all qualitative features of the nuclear strong force.

Cosmic Microwave Background (CMB)

1. CMB Origin as Medium Phase-Transition Radiation

In the AM Model, the CMB arises from the latent-heat release during Medium-to-matter crystallization. The baseline emission condition is:

- ΔE = P_Λ · ΔV
- where P_Λ is the residual Medium pressure and ΔV is expansion volume during re-seeding.

2. Effective Temperature of CMB

The equilibrium radiation temperature emerges from pressure–density ratio:

- T_CMB ∝ (P_Λ / ρ_AM)^{1/4}
- This produces the ≈2.7 K baseline observed value.

3. CMB Frequency from Medium Oscillations

CMB oscillations correspond to the fundamental vibrational mode of the Absolute Medium:

- $\nu = c / L_{fund}$
- where L_{fund} is the Fundamental Length of the Energetic Net.

4. Redshift of the CMB Through Medium Expansion

As the Medium expands, photon wavelength stretches according to:

- $\lambda(t) = \lambda_0 \cdot (R(t) / R_0)$
- where R(t) is the expansion scale factor driven by P_Λ.

5. Uniformity Condition

Uniform CMB temperature follows from extremely high AM stiffness:

- $c = \sqrt{K_{AM} / \rho_{AM}}$
- Large K_{AM} ensures rapid equalization of pressure disturbances, yielding near-perfect isotropy.

6. Origin of Anisotropies

Small CMB variations ($\sim 10^{-5}$) arise from AM density gradients during early re-seeding:

- $\delta T/T \propto \delta\rho_{AM} / \rho_{AM}$

7. Physical Summary

The CMB is interpreted as latent-heat radiation generated by the phase transition of the Absolute Medium during re-seeding. Its temperature, spectrum, and uniformity follow directly from Medium stiffness, residual pressure, and expansion dynamics, providing a mechanical, non-inflationary explanation.

Neutron-Star Saturation & Coupling Lock

1. Density-Driven Rise of Volumetric Coupling

In the AM Model, the Volumetric Coupling Coefficient sharply increases as matter approaches neutron-star density. The scaling relation is:

- C_v(ρ_m) = (κ / L_fund) · f(ρ_m)
- with f(ρ_m) → ∞ as ρ_m → ρ_sat, where ρ_sat is the saturation density of compressed matter.

2. Choked Flow: Medium Infiltration Limit

As internal porosity collapses, the Medium can no longer flow through the matter volume. The attenuation law becomes:

- Q_out = Q_in · exp(− C_v · L)
- In the neutron-star limit: C_v → ∞ ⇒ Q_out → 0
- indicating complete flow choking.

3. Formation of the Surface-Locked Interface

When AM flow is blocked internally, interaction shifts entirely to the surface. Force transmission becomes:

- F_surface = C_v · P_D · A_surface
- where P_D = k $(\Delta\rho)^2$ is Dissociative Pressure.

4. NLEP and Extreme Compression

From Non-Linear Elasticity ($\sigma = k\,\varepsilon^2$) and $\varepsilon \propto M/r^2$, pressure becomes:

- $P(r) \propto M^2 / r^4$
- At neutron-star radii ($r \to R\_NS$), $P(r)$ approaches $\sigma\_max = k\ \varepsilon\_max^2$, the Medium's elastic saturation.

5. Coupling Lock: Full AM–Matter Engagement

Coupling lock occurs when the Medium's stiffness dominates motion:

- $C\_v \to \infty \Rightarrow F\_total \approx C\_v \cdot P\_D$
- The matter becomes "welded" to the Medium, enabling ultra-fast rotations of neutron stars.

6. Gravity at Saturation (SDL limit)

Using SDL ($\rho\_AM \propto M^2/R^5$), neutron-star compaction yields:

- $\rho\_AM(R\_NS) \propto M^2 / R\_NS^5$
- This produces extreme curvature and large gravitational acceleration:
- $g\_NS \propto M^2 / R\_NS^4$

7. Summary

Neutron-star saturation occurs when matter density forces the Volumetric Coupling to diverge and the Medium's non-linear elasticity reaches its limit. Flow choking, surface locking, extreme r^{-4} pressure, and SDL-enhanced AM density combine to create the most intense non-singular gravitational environment in the cosmos.

Dissociative Limit → Pure Sink (Black-Hole Core Physics)

1. Dissociative Limit: Collapse of Material Structure

The Dissociative Limit occurs when matter density forces the Absolute Medium (AM) to exceed its elastic support. The collapse criterion is:

- $P\_D(r) \geq \sigma\_max$
- where $P\_D = k(\Delta\rho)^2$ is Dissociative Pressure and $\sigma\_max = k\ \varepsilon\_max^2$ is the maximum elastic stress of the Medium.

2. r^{-4} Divergence of Pressure at the Collapse Radius

From NLEP and displacement $\varepsilon \propto M/r^2$, the pressure profile near the core is:

- $P(r) = k(M^2 / r^4)$
- As $r \to r\_crit$, $P(r)$ diverges until it surpasses σ_max, breaking nuclear structure.

3. Breakdown of Elasticity (AM Structural Failure)

The Medium loses its ability to support displacement when:

- $\varepsilon \to \varepsilon\_max$
- At this point, $\sigma \to \sigma\_max$ and the Medium switches from elastic to purely advective (flow-dominated) behavior.

4. Pure Sink: Infinite Coupling and Unlimited Inflow

Once elastic support fails, matter dissolves into AM and a Pure Sink forms. The conditions are:

- $C_v \to \infty$
- $v_r(r) \to \infty$
- where C_v is the Volumetric Coupling coefficient and v_r is the radial inflow velocity.
- The Pure Sink becomes a region where AM inflow rate is theoretically unbounded.

5. Event Horizon: c-Limited Flow Boundary

The Event Horizon forms where inward Medium flow speed equals wave propagation speed:

- $v_r(r_H) = c$
- with $c = \text{sqrt}(K_{AM} / \rho_{AM})$. Inside r_H, no Medium ripple (light) can propagate outward.

6. Pressure-Driven Exhaust (Jet Formation)

Conservation of Medium flow requires:

- $Q_{in} = Q_{out}$

The extreme compression at the sink generates exhaust jets with velocity:

- $v_{jet} = \text{sqrt}(2(P_{core} - P_{env})/\rho_{AM})$

7. Unified Dissociation → Pure Sink Mapping

The full transition is governed by:

- Pure_Sink = f(P_D, σ_max, C_v, v_r)
- Expanded:
- Pure_Sink ∝ $(M^2 / r^4 - \sigma\_max) \cdot \exp(C\_v\ L) \cdot v\_r(r)$

8. Summary

As matter density increases, Dissociative Pressure overwhelms Medium elasticity, leading to structural collapse. The Medium transitions into a purely advective state, forming a Pure Sink with infinite coupling and inward flow. This constitutes the core physics of Black-Hole behavior in the AM Model.

Galactic Gearbox (Dark-Matter-Equivalent) from SDL + VCP

1. The Galactic Gearbox Concept

In the AM Model, flat galactic rotation curves arise naturally from a mechanical interaction between the Absolute Medium (AM) and baryonic mass. This effect is termed the Galactic Gearbox. The gearbox is produced by two coupled laws:

- • Structural Density Law (SDL): ρ_AM(R) ∝ M^2 / R^5
- • Volumetric Coupling Principle (VCP): F_eff = C_v · P_D

2. SDL-Driven Medium Density Enhancement

SDL controls how strongly the AM is compressed around baryonic mass. Evaluated at galactic radius R_gal:

- ρ_AM(R_gal) = C_SDL (M^2 / R_gal^5)
- This creates an extended AM density halo—functionally identical to dark matter.

3. Force Transmission via VCP

The AM's ability to push matter depends on volumetric coupling:

- C_v = κ / L_fund · f(ρ_m)

At galactic scales, f(ρ_m) is small but non-zero, giving:

- F_eff = C_v · P_D

4. Dissociative Pressure Driving Galactic Inflow

Dissociative Pressure decreases radially as:

- P_D(r) $\propto$ M^2 / r^4

The gradient determines inflow velocity:

- a(r) = − dP_D/dr $\propto$ M^2 / r^5

5. Galactic Gearbox Equation

The rotation velocity is set by the balance of AM push and centripetal requirement:

- v^2 / r = F_eff / m

Substituting SDL and VCP yields:

- $v^2 \propto (C\_v \cdot M^2 / r^4) \cdot r$
- Thus:
- v ∝ M / r^{3/2}
- At large r, C_v(r) increases slightly due to AM-density halo, giving:
- v ≈ const
- This reproduces flat rotation curves without dark matter.

6. Gear Ratio Interpretation

Define gear ratio as inflow coupling to tangential motion:

- G(r) = C_v(r) · ρ_AM(r)
- Using SDL: ρ_AM ∝ M^2/R^5, we obtain:
- G(r) ∝ C_v(r) · (M^2 / r^5)

7. Physical Summary

The Galactic Gearbox emerges from the combined action of SDL (which generates an AM density halo) and VCP (which determines how efficiently galactic matter is pushed by the Medium). The resulting torque-like coupling produces flat rotation curves and mimics dark matter without invoking non-baryonic mass.

Galaxy Cluster Binding Without Dark Matter (AM SDL Extension)

1. Cluster-Scale Structural Density Law (SDL) Extension

On megaparsec scales, the Absolute Medium (AM) density around a cluster of total baryonic mass M_cl and effective radius R_cl follows an SDL-like scaling with a cluster geometry factor χ_cl:

- $\rho\_AM,cl(R\_cl) = C\_SDL \cdot \chi\_cl \cdot M\_cl^2 / R\_cl^5$
- χ_cl (> 0) encodes non-sphericity, substructure, and filamentary feeding; C_SDL is the planetary-to-cosmic calibration constant.

2. Intra-Cluster Medium (ICM) Volumetric Coupling

The gas-rich ICM enhances Medium–matter coupling. With gas density ρ_g and characteristic path L_ICM, the effective coupling is:

- $C\_v,cl = (\kappa / L\_fund) \cdot f(\rho\_g)$, $Q\_out = Q\_in \cdot \exp(- C\_v,cl \cdot L\_ICM)$
- As $f(\rho\_g)$ increases, inflow is attenuated internally and the cluster "self-locks" to the AM sheath.

3. Binding from AM Dissociative Pressure Gradients

Using NLEP with $\varepsilon \propto M/r^2$, the AM pressure around the cluster core is:

- $P\_D(r) = k \cdot (M\_cl^2 / r^4)$

The inward AM push on a test mass m at radius r is:

- $a\_AM(r) = (C\_v,cl \cdot P\_D(r)) / \rho\_eff = (C\_v,cl \cdot k \cdot M\_cl^2) / (\rho\_eff \cdot r^4)$
- Here ρ_eff accounts for porous stellar+gas matter and partial shielding; higher ICM density raises C_v,cl and lowers ρ_eff via volumetric engagement.

4. Hydrostatic Equilibrium in the AM Sheath

For an isothermal ICM with temperature T and number density n, hydrostatic balance under AM push reads:

- $dP_{ICM}/dr = -\rho_g \cdot a_{AM}(r)$, with $P_{ICM} = n k_B T$

Substituting $a_{AM}(r)$ gives the cluster-scale equilibrium ODE:

- $d(nk_B T)/dr = -\rho_g \cdot (C_{v,cl} \cdot k \cdot M_{cl}^2)/(\rho_{eff} \cdot r^4)$
- Integrating yields pressure and density profiles that mimic "dark halo" support without invoking non-baryonic mass.

5. Galaxy Velocity Dispersion from AM Field

Assuming spherical shells for clarity, the line-of-sight velocity dispersion $\sigma_{los}(r)$ under AM acceleration is:

- $\sigma_{los}^2(r) \approx \int_r^\infty a_{AM}(R)\, dR \propto \int_r^\infty (M_{cl}^2 / R^4)\, dR \propto M_{cl}^2 / r^3$
- A slowly varying $C_{v,cl}(r)$ flattens $\sigma_{los}(r)$ across large radii, reproducing observationally "extended mass" behavior.

6. Gravitational Lensing from AM Density Gradients

Light follows gradients in AM density (effective optical index n_{eff}). With $n_{eff}(r) \approx 1 + A \cdot \rho_{AM,cl}(r)$, the deflection angle at impact parameter b is:

- $\alpha(b) \approx \int (\partial n_{eff}/\partial \perp)\, dz \propto \int (\partial \rho_{AM,cl}/\partial \perp)\, dz \propto \int \partial/\partial \perp [M_{cl}^2 / (R^5)]\, dz$
- Since $\rho_{AM,cl} \propto M_{cl}^2/R^5$ extends beyond baryons, the integral yields lensing equivalent to a dark-matter-like halo.

7. Binding Criterion Without Dark Matter

A cluster is bound when AM pressure support exceeds kinetic dispersion of member galaxies:

- $C\_v,cl \cdot P\_D(r) \geq \rho\_eff \cdot (v^2/r)$
- Equivalently, using the ODE in §4 and observed n(r), T(r), one solves for M_cl that satisfies binding without extra mass species.

8. External Confinement from Large-Scale AM Pressure

SDL predicts a background AM pressure P_Λ plus filament-fed density ρ_fil. The outer boundary condition becomes:

- $P\_ICM(R\_out) = P\_\Lambda + A\_fil \cdot \rho\_fil$
- This external "Medium wall" reduces evaporation losses and tightens the binding without invoking dark matter.

9. Physical Summary

The AM SDL extension and ICM-enhanced VCP generate an extended AM density halo and strong inward Medium push, providing hydrostatic support, near-flat velocity dispersions, and lensing strengths associated with galaxy clusters—without non-baryonic dark matter. Binding arises from r^{-4} pressure gradients, volumetric coupling, and external AM confinement at cluster edges.

Quantum–Classical Transition via AM Saturation Layer

1. Definition of the AM Saturation Layer

The Saturation Layer is the scale boundary where the Absolute Medium (AM) transitions from quantum-like elastic behavior to classical hydrodynamic flow. The defining criterion is:

- $\varepsilon(r) = \varepsilon\_sat$
- where $\varepsilon(r) \propto M/r^2$ is displacement strain and ε_sat is the Medium's saturation strain.

2. NLEP Governing the Transition

The Non-Linear Elasticity Principle (NLEP) controls behavior for $r \leq r\_sat$. Elastic stress follows:

- $\sigma(r) = k\,\varepsilon(r)^2 = k\,(\alpha M/r^2)^2 = k\,\alpha^2 M^2 / r^4$
- At $r = r\_sat$:
- $\sigma\_sat = k\,\varepsilon\_sat^2$

3. Surge of Volumetric Coupling at Saturation

The transition is accompanied by a sharp rise in Volumetric Coupling:

- $C\_v(r) = (\kappa/L\_fund) \cdot f(\rho\_m)$
- with $f(\rho\_m) \to \infty$ as $\rho\_m \to \rho\_sat$.
- Thus at $r = r\_sat$:
- $C\_v \to C\_{v,sat} \gg 1$

4. Quantum-Like Regime for r < r_sat

The pressure landscape is dominated by NLEP elastic stresses:

- $P(r) = k\,\alpha^2\,M^2 / r^4$
- The effective force is strongly non-linear:
- $F\_Q(r) = -\,dP/dr = 4k\,\alpha^2\,M^2 / r^5$
- This steep r^{-5} behavior produces confinement-like effects and discrete modes (quantum levels).

5. Classical Regime for r > r_sat

When r > r_sat, elastic saturation suppresses NLEP contribution and hydrodynamic flow dominates:

- $v\_r(r) = -\,Q / (2\pi r)$
- $v\_\theta(r) = \Gamma / (2\pi r)$
- These produce classical gravitational motion and macroscopic trajectories.

6. Quantum → Classical Transition Equation

The transition radius r_sat solves:

- $k\,\alpha^2\,M^2 / r\_sat^4 = \sigma\_sat$
- Thus:
- $r\_sat = (\alpha M / \varepsilon\_sat)$^{1/2}

7. Decoherence as Saturation-Driven Damping

The Medium damps short-scale oscillations at the saturation boundary. Damping coefficient:

- $\gamma\_sat \propto C\_v{,}sat \cdot \sigma\_sat$

- Large C_v,sat converts quantum-like oscillations into classical trajectories.

8. Unified Force Expression Across the Quantum–Classical Boundary

Matching the two regimes yields the piecewise force law:

- $F(r) = \{ 4k \alpha^2 M^2 / r^5 , r < r\_sat$
- $C\_v(r) P\_D(r) / \rho\_eff , r > r\_sat \}$

9. Physical Summary

The Absolute Medium produces a natural quantum–classical boundary at the Saturation Layer, where elastic non-linearity gives way to hydrodynamic flow. Inside the boundary, steep r^{-4} and r^{-5} elastic fields generate quantum-like confinement. Outside, volumetric coupling and flow dominate, producing classical trajectories. Decoherence emerges as a mechanical saturation-driven damping effect.

AM Field Tensor

1. Motivation for a Relativized AM Field Tensor

In the AM Model, spacetime curvature is replaced by Medium deformation and flow. To express this in tensor form, we construct a field tensor combining:

- • Elastic (NLEP) contributions
- • Flow (hydrodynamic) contributions
- • Density-gradient (SDL) contributions

- • Coupling (VCP) contributions

2. Displacement 4-Gradient

Let ε be displacement of the AM Energetic Net. The 4-gradient is:

- $\varepsilon\_\mu = \partial\_\mu \varepsilon$
- This plays the role analogous to metric perturbations in GR but remains Medium-based.

3. Elastic Contribution (NLEP Tensor)

Non-Linear Elasticity gives stress: $\sigma = k\,\varepsilon^2$. The elastic tensor is:

- $E\_\{\mu\nu\} = k\,\varepsilon\_\mu\,\varepsilon\_\nu$
- This replaces curvature terms ($R\_\{\mu\nu\}$) in GR.

4. Flow Tensor (Hydrodynamic Component)

Let v_μ be the 4-velocity of AM flow. The antisymmetric flow tensor is:

- $F\_\{\mu\nu\} = \partial\_\mu\, v\_\nu - \partial\_\nu\, v\_\mu$
- This captures vortex and sink motion of the AM.

5. Density-Gradient Tensor (SDL Contribution)

The Structural Density Law gives $\rho\_AM \propto M^2/R^5$. Its relativized gradient contribution is:

- $S\_\{\mu\nu\} = \partial\_\mu\, \rho\_AM\, \partial\_\nu\, \rho\_AM$

6. Coupling Tensor (VCP Contribution)

Volumetric Coupling depends on density. The coupling tensor is:

- $C_{\mu\nu} = C_v \cdot E_{\mu\nu}$
- with $C_v = \kappa/L_{fund} \cdot f(\rho_m)$.

7. Unified AM Field Tensor

The unified AM Field Tensor combines the above components:

- $T_{\mu\nu}^{AM} = E_{\mu\nu} + F_{\mu\nu} + S_{\mu\nu} + C_{\mu\nu}$
- Explicitly:
- $T_{\mu\nu}^{AM} = k\, \varepsilon_\mu\, \varepsilon_\nu + (\partial_\mu v_\nu - \partial_\nu v_\mu) + (\partial_\mu \rho_{AM}\, \partial_\nu \rho_{AM}) + C_v\, k\, \varepsilon_\mu\, \varepsilon_\nu$

8. Relativized AM Field Equation

The AM Field Equation equates divergence of the AM tensor to effective force density:

- $\partial^\mu T_{\mu\nu}^{AM} = f_\nu$
- where f_ν includes dissociative pressure gradients ($-\partial_\nu P_D$) and inertial flow ($\rho_{AM}\, dv_\nu/d\tau$).

9. Physical Summary

The relativized AM Field Tensor provides a unified representation of elasticity, flow, density gradients, and volumetric coupling. It replaces curvature tensors in GR with mechanical Medium interactions, preserving relativistic structure while remaining rooted in the Absolute Medium framework.

AM Stress–Energy Tensor for Cosmology (Λ_{AM})

1. Definition of the AM Stress–Energy Tensor

In the AM Model, the stress–energy content of the Universe arises from mechanical properties of the Absolute Medium (AM), not spacetime curvature. The AM stress–energy tensor is defined as:

- $T_{\mu\nu}^{AM} = E_{\mu\nu} + F_{\mu\nu} + S_{\mu\nu} + C_{\mu\nu} + \Lambda_{AM}\, g_{\mu\nu}$
- where each term corresponds to elastic, flow, density-gradient, coupling, and cosmological-pressure sectors.

2. Elastic Sector from NLEP

From the Non-Linear Elasticity Principle ($\sigma = k\, \varepsilon^2$), the elastic tensor contribution is:

- $E_{\mu\nu} = k\, \varepsilon_\mu\, \varepsilon_\nu$
- representing AM stiffness and internal energy.

3. Flow Sector

The antisymmetric AM flow tensor is defined by:

- $F_{\mu\nu} = \partial_\mu v_\nu - \partial_\nu v_\mu$
- encoding sink–vortex inflow and rotational dynamics.

4. Structural Density Law (SDL) Sector

From cluster-scale and cosmic-scale density gradients:

- $S_{\mu\nu} = \partial_\mu \rho_{AM} \partial_\nu \rho_{AM}$
- with $\rho_{AM} \propto M^2 / R^5$.

5. VCP Sector

Volumetric coupling between matter and AM contributes:

- $C_{\mu\nu} = C_v \cdot E_{\mu\nu}$
- where $C_v = \kappa/L_{fund} \cdot f(\rho_m)$.

6. AM Cosmological Constant Term

The AM background pressure P_Λ defines Λ_{AM}. We introduce:

- $\Lambda_{AM} = P_\Lambda / K_{AM}$
- with K_{AM} the baseline Medium stiffness.
- The cosmological term enters the tensor as:
- $\Lambda_{AM} g_{\mu\nu}$
- where $g_{\mu\nu}$ is the flat-background metric (AM reference frame).

7. Unified Cosmological AM Field Equation

The cosmological field equation equates the divergence of the AM tensor to zero (Medium conservation):

- $\partial^\mu T_{\mu\nu}^{AM} = 0$
- Expanding yields the cosmological evolution equation:
- $\partial^\mu(E_{\mu\nu} + F_{\mu\nu} + S_{\mu\nu} + C_{\mu\nu}) = -\partial_\nu (\Lambda_{AM})$

8. Expansion from AM Pressure

Using the AM cosmological constant term, the expansion rate obeys:

- $(\dot{R}/R)^2 \propto \Lambda\_AM = P\_\Lambda / K\_AM$
- Thus cosmic expansion is pressure-driven rather than curvature-driven.

9. Physical Summary

The AM stress–energy tensor unifies elastic, flow, density, and coupling sectors of the Absolute Medium into a relativized framework. The cosmological term Λ_AM emerges from the Medium's intrinsic pressure, producing expansion without invoking vacuum energy or spacetime curvature. This tensor replaces the Einstein tensor as the governing dynamical object for cosmology in the AM Model.

AM Hamiltonian & Lagrangian Field Formulation

1. Field Variable and Action Principle

The fundamental dynamical variable of the AM Model is the Medium displacement field:

- $\varepsilon = \varepsilon(x, t)$
- Dynamics follow from an action principle with Lagrangian density ⯑_AM(ε, ∂ε).

2. AM Lagrangian Density

The AM Lagrangian density consists of kinetic, elastic, nonlinear (NLEP), and coupling (VCP) terms:

- ⯑_AM = $\frac{1}{2}$ ρ_AM $(\partial\varepsilon/\partial t)^2$ – $\frac{1}{2}$ K_AM $(\nabla\varepsilon)^2$ – $k\,\varepsilon^4$ + C_v $(\Delta\rho)\,\varepsilon$
- • First term: kinetic energy of Medium oscillation.

- • Second term: elastic spatial energy from stiffness K_AM.
- • Third term: NLEP nonlinear self-interaction.
- • Fourth term: coupling of matter to AM.

3. Euler–Lagrange Equation → AM Wave Equation

Applying the Euler–Lagrange operator:

- $\partial ◈/\partial \varepsilon - d/dt\,(\partial ◈/\partial \dot{\varepsilon}) - \nabla\cdot(\partial ◈/\partial(\nabla \varepsilon)) = 0$

yields the relativistic AM field equation:

- $\rho\_AM\ \ddot{\varepsilon} - K\_AM\ \nabla^2\varepsilon + 4k\ \varepsilon^3 = C\_v\ (\Delta\rho)$

4. Conjugate Momentum Density

The momentum conjugate to ε is:

- $\pi = \partial ◈\_AM\ /\ \partial\dot{\varepsilon} = \rho\_AM\ \dot{\varepsilon}$
- representing Medium kinetic storage.

5. Hamiltonian Density

The Hamiltonian density is defined as:

- $◈\_AM = \pi\,\dot{\varepsilon} - ◈\_AM$

Substituting $\pi = \rho\_AM\ \dot{\varepsilon}$ gives:

- $◈\_AM = ½\ \rho\_AM\ \dot{\varepsilon}^2 + ½\ K\_AM\ (\nabla\varepsilon)^2 + k\ \varepsilon^4 - C\_v\ (\Delta\rho)\ \varepsilon$

6. Hamilton's Equations in AM Form

Hamilton's field equations become:

- $\dot{\varepsilon}$ = ∂�_AM / ∂π = π / ρ_AM
- $\dot{\pi}$ = − ∂�_AM / ∂ε = K_AM $\nabla^2 \varepsilon$ − 4k ε^3 + C_v (Δρ)
- Combining these yields the full AM evolution PDE.

7. Energy–Momentum Tensor from �_AM

The AM stress–energy tensor follows from the Lagrangian via:

- T_{μν}^{AM} = ∂_μ ε ∂_ν ε − g_{μν} �_AM
- This tensor governs cosmic evolution (Λ_AM sector).

8. Physical Summary

The AM Lagrangian–Hamiltonian framework reveals a unified Medium dynamics: linear wave motion, nonlinear NLEP confinement, matter–Medium coupling, and cosmic-scale stress–energy behavior all arise from �_AM. This positions the AM Model as a full field theory with explicit mechanical foundations.

Relativistic AM Wave Equation (Hyperbolic PDE for ε)

1. AM Displacement Field ε(x, t)

In the AM Model, all waves—including photons—are transverse oscillations of the Absolute Medium. The fundamental field variable is the displacement:

- ε = ε(x, t)
- representing deviation of the Energetic Net from equilibrium.

2. Relativistic Stiffness Condition

The Medium's wave speed is determined by its stiffness-to-density ratio:

- $c^2 = K\_AM / \rho\_AM$
- This is inserted directly into the PDE structure.

3. Hyperbolic Wave Operator

The relativistic d'Alembertian operator for AM excitations is:

- $\square = \partial^2/\partial t^2 - c^2 \nabla^2$
- where $c = sqrt(K\_AM/\rho\_AM)$.

4. NLEP Contribution to AM Wave Dynamics

Non-Linear Elasticity adds a quadratic correction term to AM dynamics:

- $\sigma = k\, \varepsilon^2$
- Variation of σ contributes a nonlinear PDE term:
- $N(\varepsilon) = \partial\sigma/\partial\varepsilon = 2k\, \varepsilon$

5. VCP Source Term

Matter couples to the Medium volumetrically. The relativistic source term is:

- $J(x,t) = C\_v \cdot (\Delta\rho)$
- where $\Delta\rho$ is density deficit created by matter.

6. Full AM Hyperbolic PDE

Combining the linear wave operator, NLEP nonlinearity, and VCP source yields the master equation:

- $\Box\, \varepsilon + 2k\, \varepsilon = J(x, t)$
- Explicitly:
- $\partial^2\varepsilon/\partial t^2 - c^2\, \nabla^2\, \varepsilon + 2k\, \varepsilon = C\_v\, (\Delta\rho)$

7. Photon as a Relativistic AM Ripple

Setting $\Delta\rho = 0$ gives free-wave solutions. A photon is:

- $\varepsilon\_photon = A \exp[i(k{\cdot}x - \omega t)]$
- with dispersion relation:
- $\omega^2 = c^2\, k^2$

8. Massive Excitations from Nonlinear Term

The nonlinear term $2k\varepsilon$ produces mass-like behavior for bound oscillatory modes:

- $\omega^2 = c^2\, k^2 + 2k$
- corresponding to confined, non-radiative excitations.

9. Physical Summary

The relativistic AM wave equation is a nonlinear hyperbolic PDE combining stiffness-controlled propagation, elastic nonlinearity, and matter coupling. Light, forces, and quantum transitions emerge from different regimes of this unified Medium equation.

Superposition as Overlapping AM Pressure Modes

1. Fundamental Pressure Mode Definition

In the AM Model, all wave-like behavior arises from oscillations of the Absolute Medium (AM). Each mode is defined by a pressure oscillation field:

- $P_i(x,t) = k \cdot \varepsilon_i(x,t)^2$
- where ε_i is the displacement amplitude of the i-th mode and k is the NLEP elastic constant.

2. Linear Superposition of Displacements

Displacement fields superpose linearly because ε is a geometric offset of the Energetic Net:

- $\varepsilon_{total}(x,t) = \Sigma\, \varepsilon_i(x,t)$
- This is the primary superposition rule at the displacement level.

3. Nonlinear Pressure Superposition (NLEP)

Because pressure follows NLEP ($\sigma = k\varepsilon^2$), combined pressure is nonlinear:

- $P_{total} = k \cdot (\varepsilon_{total})^2 = k\, (\Sigma\, \varepsilon_i)^2$
- Expanding:
- $P_{total} = k\, \Sigma\, \varepsilon_i^2 + 2k\, \Sigma_{i<j}\, \varepsilon_i\, \varepsilon_j$
- Cross-terms ($\varepsilon_i\, \varepsilon_j$) generate interference.

4. Constructive and Destructive Interference

If two modes ε_1 and ε_2 overlap, the interference pressure term is:

- $P_{int} = 2k \varepsilon_1 \varepsilon_2$
- • Positive $\varepsilon_1 \varepsilon_2$ → constructive interference.
- • Negative $\varepsilon_1 \varepsilon_2$ → destructive interference.

5. Superposition of Photon Modes

A photon is a transverse AM displacement:

- $\varepsilon_{ph} = A \cos(k \cdot x - \omega t)$

Two photons superpose as:

- $\varepsilon_{total} = A_1 \cos(\theta_1) + A_2 \cos(\theta_2)$
- Pressure interference then becomes:
- $P_{total} = k(A_1^2 \cos^2\theta_1 + A_2^2 \cos^2\theta_2 + 2A_1A_2 \cos\theta_1 \cos\theta_2)$

6. Quantum Superposition as AM Pressure Mode Overlap

In the AM interpretation of quantum mechanics, a "superposition state" is simply the physical coexistence of overlapping displacement modes:

- $\varepsilon_{superposed} = \Sigma \varepsilon_i$
- Probability amplitudes correspond to pressure contributions from each mode.

7. Collapse as Nonlinear Saturation

When overlapping modes reach the AM Saturation Layer (ε → ε_sat), nonlinear amplification forces mode selection:

- P_total → k ε_sat^2
- This suppresses cross-terms and yields a classical outcome.

8. Physical Summary

Superposition in the AM Model is a mechanical overlap of pressure modes. Displacements add linearly but pressures interfere nonlinearly, producing constructive and destructive patterns. Quantum superposition emerges as multi-mode displacement coexistence, while collapse occurs when nonlinear elastic saturation forces the Medium into a single-mode configuration.

AM Canonical Quantization (ε, π → operators)

1. Classical Field Variables (ε, π)

The AM Model uses the Medium displacement field ε(x,t) as the fundamental classical variable, with conjugate momentum:

- ε = ε(x,t)
- π = ρ_AM $\dot{\varepsilon}$
- These arise from the AM Hamiltonian–Lagrangian formulation.

2. Classical Poisson Bracket Structure

The canonical Poisson bracket between ε and π is:

- $\{ \varepsilon(x), \pi(x'')] = \delta(x - x'')$
- This is the starting point for quantization.

3. Canonical Quantization Prescription

Canonical quantization promotes fields and momenta to operators and replaces Poisson brackets with commutators:

- $\{A, B\} \to (1/i\hbar)\ [\hat{A}, \hat{B}]$

4. Fundamental AM Commutation Relations

Thus the AM displacement and momentum satisfy:

- $[\ \hat{\varepsilon}(x), \hat{\pi}(x'')\ ] = i\hbar\ \delta(x - x'')$
- while all other equal-time commutators vanish:
- $[\ \hat{\varepsilon}(x), \hat{\varepsilon}(x'')\ ] = 0$
- $[\ \hat{\pi}(x), \hat{\pi}(x'')\ ] = 0$

5. Mode Expansion of the AM Field

Expanding $\hat{\varepsilon}(x,t)$ in normal modes of the Medium:

- $\hat{\varepsilon}(x,t) = \Sigma_k \left(a_k\ u_k(x)\ e^{-i\omega_k t} + a_k^\dagger\ u_k^*(x)\ e^{+i\omega_k t} \right)$
- with dispersion relation $\omega_k^2 = c^2\ k^2$ for photon-like AM ripples.

6. Creation–Annihilation Operator Algebra

The commutation relations for a_k and $a_k^\dagger$ follow from the field algebra:

- $[\ a_k, a_q^\dagger\ ] = \delta_{kq}$

- $[a\_k, a\_q] = 0$
- These define the quantum structure of AM excitations.

7. Hamiltonian Operator for the AM Field

Quantizing the Hamiltonian density H_AM yields:

- $\hat{H} = \Sigma\_k \hbar\omega\_k (a\_k\dagger a\_k + 1/2)$
- The $1/2\hbar\omega\_k$ term originates from zero-point AM fluctuations.

8. NLEP-Induced Anharmonic Quantum Term

Because $\sigma = k\varepsilon^2$, quantization introduces a quartic operator term:

- $\hat{H}\_NLEP \propto k \int \hat{\varepsilon}^4 d^3x$
- This produces self-interaction between AM quanta, analogous to nonlinear optomechanics.

9. Physical Summary

Canonical quantization in the AM Model elevates Medium displacement and momentum to operator fields. The resulting algebra mirrors quantum field theory but remains grounded in mechanical Medium physics. AM quanta (photons, confined modes, etc.) arise naturally as excitations of $\hat{\varepsilon}$, and nonlinear NLEP terms generate self-interactions absent in linear vacuum models.

AM Path-Integral Formulation (Action over Medium Configurations)

1. Medium Configuration Space Ω[ε(x,t)]

In the AM Model, the fundamental degree of freedom is the displacement field of the Absolute Medium:

- $\varepsilon = \varepsilon(x,t)$
- A complete physical history corresponds to a trajectory ε(x,t) through configuration space Ω.

2. Action Functional for the Absolute Medium

The action is the spacetime integral of the AM Lagrangian density ⟐_AM:

- S[ε] = ∫ d^3x dt ⟐_AM(ε, ∂ε)
- Substituting ⟐_AM from the field formulation gives:
- ⟐_AM = ½ ρ_AM ε^2 – ½ K_AM $(\nabla\varepsilon)^2$ – k ε^4 + C_v (Δρ) ε

3. AM Path-Integral Propagator

The transition amplitude between Medium configurations ε_i(x) and ε_f(x) is:

- K(ε_f, ε_i) = ∫_Ω Dε(x,t) exp[(i/ħ) S[ε]]
- The functional measure Dε sums over all possible Medium histories.

4. Classical Limit from Stationary Action

In the limit ħ → 0, the integral is dominated by paths satisfying:

- $\delta S[\varepsilon] = 0$
- yielding the AM field equation:
- $\rho\_AM\, \ddot{\varepsilon} - K\_AM\, \nabla^2\varepsilon + 4k\, \varepsilon^3 = C\_v\, (\Delta\rho)$

5. Fluctuations Around Classical Paths

Write the field as $\varepsilon = \varepsilon\_cl + \eta$, where $\eta(x,t)$ is a fluctuation mode. Expanding the action to second order gives:

- $S[\varepsilon] = S[\varepsilon\_cl] + ½ \int \eta \,\backslash, O \,\backslash, \eta + ...$
- where O is the AM fluctuation operator:
- $O = \rho\_AM\, \partial^2/\partial t^2 - K\_AM\, \nabla^2 + 12k\, \varepsilon\_cl^2$

6. AM Partition Functional (Quantum Medium Corrections)

The vacuum partition functional is:

- $Z = \int D\varepsilon \exp[\, (i/\hbar)\, S[\varepsilon]\,]$
- This defines quantum corrections to AM elasticity, propagation, and coupling.

7. Photon as a Path-Integral Over Transverse AM Ripples

For $\Delta\rho = 0$, the AM action becomes quadratic in ε, giving Gaussian path integrals:

- $K\_photon = \int D\varepsilon \exp[\, (i/\hbar) \int (½\, \rho\_AM\, \dot{\varepsilon}^2 - ½\, K\_AM\, (\nabla\varepsilon)^2)\, d\text{^}3x\, dt\,]$
- Solutions correspond to all possible photon (AM ripple) paths.

8. Decoherence as Path-Integral Weight Suppression

Nonlinear NLEP and VCP terms suppress non-classical trajectories when ε approaches the Saturation Layer ε_sat:

- $\exp[\,(i/\hbar) \int -k\, \varepsilon^4\, dV\, dt\,] \to 0$
- This dynamically selects classical outcomes (measurement behavior).

9. Physical Summary

The AM path-integral formulation treats quantum evolution as a weighted sum over all Medium displacement histories. Elastic stiffness defines propagation, NLEP introduces nonlinear weighting, VCP couples matter to the Medium, and saturation leads to decoherence. This yields a unified mechanical origin for both quantum and classical behavior.

AM Renormalization: Cutoff at L_fund

1. Motivation: UV Control by Medium Grain

In the AM Model, the Energetic Net possesses a fundamental grain length L_fund. All mode sums and loop integrals are regulated by this physical cutoff, removing ultraviolet (UV) divergences by construction.

2. Spectral Cutoff at L_fund

Define the maximum admissible wavenumber and frequency as:

- $k\_max = \pi / L\_fund$
- $\omega\_max = c \cdot k\_max = c \cdot \pi / L\_fund$

- with $c = \sqrt{K_{AM} / \rho_{AM}}$.

3. Mode Density and Zero-Point Energy

For a cubic quantization volume V, the AM mode density with cutoff is:

- $g(k)\, dk = (V / 2\pi^2)\, k^2\, dk\, ,\ 0 \le k \le k_{max}$

Zero-point energy of transverse AM modes becomes:

- $E_0 = (1/2)\, \hbar \int_0^{k_{max}} \omega(k)\, g(k)\, dk = (\hbar\, c\, V / 4\pi^2) \int_0^{k_{max}} k^3\, dk$

Evaluating with cutoff:

- $E_0 = (\hbar\, c\, V / 16\pi^2)\, k_{max}^4 = (\hbar\, c\, V / 16\pi^2)\, (\pi / L_{fund})^4$

4. Renormalized Baseline Pressure P_Λ

Define the AM baseline pressure (cosmological sector) from zero-point density $\rho_0 = E_0 / V$:

- $\rho_0 = (\hbar\, c / 16\pi^2)\, (\pi / L_{fund})^4$

Map to baseline pressure via Medium equation of state:

- $P_\Lambda = \xi \cdot \rho_0$
- with ξ determined by the AM elastic sector ($\xi \approx 1$ for ultra-stiff modes).

5. Loop Regularization with Physical Cutoff

For a generic one-loop self-energy integral in the AM field theory:

- $\Sigma(0) \propto \int^{k_{max}} d^3k / (2\pi)^3 \cdot (1/2\omega_k)$
- Using $\omega_k = \sqrt{c^2 k^2 + m_{eff}^2 c^4 / \hbar^2}$ and the cutoff k_{max}, $\Sigma(0)$ is finite and scales as k_{max}^2 for large cutoff, i.e., $\Sigma(0) \propto k_{max}^2 \to (\pi / L_{fund})^2$.

6. Counterterm Calibration at L_fund

Define renormalized elastic and coupling constants at a reference scale $\mu_0 \equiv 1 / L_{fund}$:

- $k_R(\mu_0) = k_B + \delta k(L_{fund})$
- $C_v^R(\mu_0) = C_v^B + \delta C_v(L_{fund})$
- where B denotes bare parameters; counterterms δk, δC_v absorb finite L_{fund}-dependent pieces and are fixed by laboratory or astrophysical observables (e.g., c, P_Λ, planetary g).

7. Renormalization Group (RG) Flow for μ < 1/L_fund

For processes with characteristic scale $\mu \ll 1/L_{fund}$, define running couplings via RG equations:

- $\mu\, d k_R / d\mu = \beta_k(k_R, C_v^R, \ldots)$
- $\mu\, d C_v^R / d\mu = \beta_{C_v}(k_R, C_v^R, \ldots)$
- with β-functions computed from finite loop integrals capped at k_{max}.

8. Consistency with SDL and NLEP

The cutoff sets the micro-to-macro bridge:

- $C\_v \approx \kappa / L\_fund$
- and ensures the NLEP field pressure near matter remains finite:
- $P(r) = k \alpha^2 M^2 / r^4 \rightarrow$ bounded at $r \gtrsim L\_fund$

9. Observational & Experimental Windows

Finite cutoff implies testable plateaus in:

- • High-frequency dispersion ($\omega \lesssim \omega\_max$)
- • Short-baseline gravitational gradients ($r \gtrsim L\_fund$)
- • Vacuum-energy-derived P_Λ calibrated against SDL at planetary/cosmic scales.

10. Summary

AM Renormalization introduces a physical UV cutoff at L_fund, rendering mode sums and loop corrections finite, fixing counterterms at $\mu\_0 = 1/L\_fund$, and ensuring consistency with NLEP and SDL across scales. The cutoff also ties vacuum pressure P_Λ to measurable elastic parameters, closing the micro–macro loop in the AM Model.

Quantum Potential as AM Pressure Curvature Term

1. AM Displacement and Pressure Fields

The AM Model uses displacement $\varepsilon(x,t)$ as the fundamental field. Dissociative Pressure in NLEP form is:

- $P(x,t) = k\,\varepsilon(x,t)^2$
- where k is the AM elastic nonlinearity constant.

2. Pressure Curvature Definition (Quantum Potential Analogue)

Define curvature of pressure as the Laplacian:

- $\nabla^2 P = \nabla^2(k\,\varepsilon^2)$
- Expanding gives:
- $\nabla^2 P = 2k(\varepsilon\, \nabla^2 \varepsilon + |\nabla \varepsilon|^2)$
- This quantity plays the role of a quantum potential in AM mechanics.

3. AM Quantum Potential Term Q_AM

Define the AM quantum potential as the curvature of pressure normalized by AM density ρ_AM:

- $Q_{AM} = -\,(1 / \rho_{AM})\, \nabla^2 P$
- Thus:
- $Q_{AM} = -\,(2k / \rho_{AM})(\varepsilon\, \nabla^2 \varepsilon + |\nabla \varepsilon|^2)$

4. Relation to Bohm–Type Quantum Potential

In Bohmian mechanics, the quantum potential is:

- $Q_B = -\,(\hbar^2 / 2m)\, (\nabla^2 R / R)$
- The AM equivalent emerges by identifying the amplitude R with AM displacement ε:
- $R \leftrightarrow \varepsilon$
- and noting that $\nabla^2 \varepsilon / \varepsilon$ and $|\nabla \varepsilon|^2 / \varepsilon^2$ terms appear naturally in $\nabla^2(\varepsilon^2)$.

5. Effective Force from AM Quantum Potential

The AM quantum force on matter is the gradient of Q_AM:

- F_Q = − ∇Q_AM
- Substituting Q_AM yields:
- $F\_Q = \nabla [(2k / \rho\_AM)(\varepsilon \nabla^2\varepsilon + |\nabla\varepsilon|^2)]$

6. Quantum Interference from Pressure Curvature

When two AM modes ε_1 and ε_2 overlap:

- $\varepsilon\_total = \varepsilon_1 + \varepsilon_2$
- The quantum potential term becomes:
- $Q\_AM \propto -\nabla^2[(\varepsilon_1 + \varepsilon_2)^2]$
- Cross terms $2\varepsilon_1\varepsilon_2$ generate full interference structure.

7. Classical Limit (Smooth ε)

When ε varies slowly over space:

- $\nabla\varepsilon \to$ small and $\nabla^2\varepsilon \to$ small
- Thus:
- $Q\_AM \to 0$
- recovering classical mechanics automatically.

8. Physical Summary

The AM quantum potential arises as the curvature of nonlinear Medium pressure $P = k\varepsilon^2$. It generates interference, guides particle trajectories, and collapses in the classical limit. This provides a mechanical, Medium-based analogue to the quantum potential of Bohmian mechanics without introducing new postulates.

Measurement Collapse as Saturation-Layer Bifurcation

1. AM Displacement Modes in Superposition

Before measurement, the system is in a superposition of displacement modes of the Absolute Medium (AM):

- $\varepsilon\_total(x,t) = \Sigma\_i\ \varepsilon\_i(x,t)$
- These modes overlap linearly at ε but combine nonlinearly at pressure $P = k\varepsilon^2$.

2. Pressure Nonlinearity and Mode Competition

Total AM pressure is:

- $P\_total = k\ (\Sigma\_i\ \varepsilon\_i)^2 = k\ \Sigma\_i\ \varepsilon\_i^2 + 2k\ \Sigma\_{i<j}\ \varepsilon\_i\ \varepsilon\_j$
- The cross–interaction terms ($2k\ \varepsilon\_i\ \varepsilon\_j$) generate competition between modes.

3. Saturation-Layer Onset (Critical Condition)

The AM has a nonlinear saturation strain ε_sat where elastic response reaches its maximum:

- $\sigma\_max = k\ \varepsilon\_sat^2$

Collapse (measurement) begins when local displacement approaches this value:

- $|\varepsilon\_total| \to \varepsilon\_sat$

4. Bifurcation Equation for Competing Modes

Write $\varepsilon\_total = \varepsilon\_a + \varepsilon\_b$ for two competing modes. The nonlinear evolution near saturation is:

- $d\varepsilon\_a/dt = -\partial P\_total/\partial\varepsilon\_a = -2k(\varepsilon\_a + \varepsilon\_b)(1)$
- $d\varepsilon\_b/dt = -\partial P\_total/\partial\varepsilon\_b = -2k(\varepsilon\_a + \varepsilon\_b)(2)$
- Near ε_sat, these equations create a bifurcation structure where only one mode survives.

5. Mode Selection Criterion

The dominant mode is the one whose local energy density is largest:

- $E\_i = \int k\, \varepsilon\_i\hat{}2\, dV$

The bifurcation selects ε_sel such that:

- $\varepsilon\_sel = \arg\max\_i\,(E\_i)$

6. Collapse Dynamics (AM Damping Term)

When saturation activates, the AM damping coefficient rises sharply:

- $\gamma\_sat \propto C\_v\, \varepsilon\_sat\hat{}2$

Evolution of each mode obeys:

- $d\varepsilon\_i/dt = -\gamma\_sat\, \varepsilon\_i$ + nonlinear competition terms

All non-selected modes decay exponentially:

- $\varepsilon\_j \to 0$ for $j \neq sel$

7. Classical Outcome After Collapse

After the bifurcation, the Medium settles into a single-mode configuration:

- $\varepsilon_{final} \approx \varepsilon_{sel}$
- This is interpreted as the measured classical result.

8. Physical Summary

Measurement collapse in the AM Model arises from nonlinear NLEP dynamics and the Medium's saturation threshold. As ε_{total} approaches ε_{sat}, competing modes undergo nonlinear bifurcation, damping eliminates all but the dominant mode, and the system transitions from a superposed quantum regime to a single classical outcome.

Quantum Tunneling as Local Pressure Undercut

1. AM Displacement Field and Pressure Landscape

Quantum tunneling in the AM Model arises from nonlinear deformation of the Absolute Medium (AM). The Medium displacement field is:

- $\varepsilon(x,t)$
- and the associated NLEP pressure is:
- $P(x) = k\,\varepsilon(x)^2$

2. Potential Barrier as Pressure Wall

A classical potential barrier corresponds to a region where AM pressure grows sufficiently large. For a barrier centered at $x=0$:

- $P_{bar}(x) = k\,\varepsilon_{bar}(x)^2$
- with ε_{bar} representing the deformation induced by the barrier material.

3. Local Pressure Undercut — Tunneling Trigger

For tunneling to occur, the total pressure landscape must be locally reduced beneath the barrier by interference or Medium fluctuation:

- $P_{total}(x) = k\,(\varepsilon_{bar}(x) + \varepsilon_{wave}(x))^2$
- An undercut occurs when:
- $P_{total}(x) < P_{bar}(x)$
- This is possible because displacement superposition is linear but pressure is nonlinear.

4. Pressure-Gradient Criterion for Penetration

Define the effective force on a particle as the AM pressure gradient:

- $F(x) = -\nabla P_{total}(x)$
- Tunneling requires a region x_{tunnel} where the net pressure gradient reverses sign:
- $\nabla P_{total}(x_{tunnel}) < 0$
- i.e., the barrier "pulls" inward due to local undercut.

5. Interference-Driven Undercut Equation

Let ε_wave be a mode approaching the barrier. The interference term is:

- $P_{int} = 2k\, \varepsilon_{bar}\, \varepsilon_{wave}$
- Tunneling occurs if this term is sufficiently negative:
- $2k\, \varepsilon_{bar}\, \varepsilon_{wave} < 0$ and $|2\, \varepsilon_{bar}\, \varepsilon_{wave}| > \varepsilon_{bar}^2$
- giving net pressure cancellation inside the barrier.

6. Undercut from AM Grain-Scale Fluctuations

Zero-point AM fluctuations at scale L_fund produce effective displacement noise $\eta(x)$:

- $\varepsilon(x) = \varepsilon_{bar}(x) + \eta(x)$
- Undercut condition becomes:
- $P(x) = k\, (\varepsilon_{bar} + \eta)^2 < k\, \varepsilon_{bar}^2$
- which reduces to:
- $2\, \varepsilon_{bar}\, \eta + \eta^2 < 0$

7. AM Transmission Coefficient

Probability of tunneling corresponds to integrated weight of undercut-enabled paths:

- $T \propto \int \exp[\, - (1/\hbar) \int |\nabla P_{total}|\, dx\,]\, D\varepsilon$
- Directly analogous to path-integral tunneling, but in AM it is fully mechanical.

8. Stabilization After Barrier Crossing

Once past the barrier, displacement returns to free-wave oscillation:

- $\varepsilon(x) \to A \cos(kx - \omega t)$
- The Medium reestablishes the free pressure profile $P = k\, \varepsilon^2$.

9. Summary

Quantum tunneling in the AM Model arises from nonlinear cancellation (undercut) of local Medium pressure caused by interference or grain-scale fluctuations. When the AM pressure landscape dips beneath the barrier profile, the effective force reverses sign and allows penetration. This provides a mechanical explanation of tunneling without invoking probabilistic postulates.

Coherent Tunneling Flows Through AM Barriers

1. AM Barrier & Coherent Inlet State

Consider a coherent AM–matter stream incident on a finite barrier region of width d. The inlet displacement is:

- $\varepsilon_{in}(x,t) = A_{in} \cos(kx - \omega t)$

The barrier is modeled as a raised pressure profile due to matter stiffness/density contrast:

- $P_{bar}(x) = k\, \varepsilon_{bar}(x)^2, x \in [0, d]$

2. AM Hyperbolic Field Equation with Nonlinearity

Inside and around the barrier the displacement obeys:

- $\partial^2\varepsilon/\partial t^2 - c^2\, \nabla^2\varepsilon + 2k\, \varepsilon = C_v(x)\, (\Delta\rho(x))$

- where $c = \sqrt{(K\_AM / \rho\_AM)}$, and C_v encodes the local Volumetric Coupling.

3. Coherent Mode Decomposition (Left/ Barrier/Right)

Assuming a quasi-1D geometry:

- Region L (x < 0): ε_L = A_in e^{i(kx−ωt)} + A_ref e^{−i(kx+ωt)}
- Barrier (0 ≤ x ≤ d): ε_B = B_+ e^{κx} e^{−iωt} + B_- e^{−κx} e^{−iωt}
- Region R (x > d): ε_R = A_tr e^{i(kx−ωt)}
- with $\kappa^2 = (\omega\_0^2 - \omega^2)/c^2$, $\omega\_0 = \sqrt{(2k)}$ (effective mass gap from NLEP).

4. Boundary Matching & Coherent Flux

At x = 0 and x = d:

- ε_L|_{0−} = ε_B|_{0+} , ∂xε_L|_{0−} = ∂xε_B|_{0+}
- ε_B|_{d−} = ε_R|_{d+} , ∂xε_B|_{d−} = ∂xε_R|_{d+}

Define coherent AM flux density (analogue of probability current):

- J = (K_AM/ρ_AM) · Im(ε* ∂xε)
- Flux conservation demands J_in = J_ref + J_tr (neglecting tiny γ_AM losses).

5. Transmission Coefficient (Coherent Tunneling Flow)

Solving the boundary system gives the transmission amplitude:

- $T\_amp = A\_tr / A\_in = [\ 1 + ((\kappa^2 + k^2)^2 / (4 k^2 \kappa^2))\ \sinh^2(\kappa\ d)\ ]^{-1/2}$

Transmission probability (coherent flow fraction):

- $T = |T\_amp|^2 = [\ 1 + ((\kappa^2 + k^2)^2 / (4 k^2 \kappa^2))\ \sinh^2(\kappa\ d)\ ]^{-1}$

6. Nonlinear Undercut & Phase-Locked Boost

When a co-propagating support mode ε_sup overlaps the barrier, the effective pressure lowers:

- $P\_eff = k\ (\varepsilon\_bar + \varepsilon\_sup)^2 < k\ \varepsilon\_bar^2$ if $2\ \varepsilon\_bar\ \varepsilon\_sup + \varepsilon\_sup^2 < 0$
- This reduces κ and increases T exponentially via $\sinh^2(\kappa\ d)$.

7. AC Pressure Bias & Resonant Flow

Applying an oscillatory pressure bias $\Delta P(t)$ across the barrier yields time-dependent phase:

- $d\Delta\theta/dt = \Delta P(t)/\rho\_eff$
- Near resonance ($\omega \approx \omega\_0$), coherent sidebands open and T is enhanced (AM analogue of resonant tunneling).

8. Saturation-Layer Crossover

- As $|\varepsilon| \to \varepsilon\_sat$ within the barrier, NLEP drives mode selection and clamps phase fluctuations, yielding a stabilized transmitted mode with reduced interference (quantum→classical crossover).

9. Summary

In the AM Model, coherent tunneling flows arise when phase-coherent AM displacement penetrates finite pressure barriers. The transmission is governed by boundary-matched hyperbolic solutions with an NLEP mass gap, enhanced by nonlinear undercut and AC bias. Saturation transitions restore classical behavior beyond the barrier.

AM Superlattices & Multi-Barrier Resonant Tunneling

1. Periodic AM Barrier Model

- Consider an AM superlattice of N identical cells, each with period $a = d + s$, where d is barrier width and s is well spacing. The raised AM pressure in each barrier segment is $P_{bar}(x) = k \cdot \varepsilon_{bar}(x)^2$; wells are near-free regions ($\Delta\rho \approx 0$) where the linear AM wave equation holds.

2. Cell Transfer Matrices (Quasi-Linear Regime)

- Using a monochromatic ansatz $\varepsilon(x,t) = \mathrm{Re}\{ \psi(x) \cdot e^{(-i \omega t)} \}$, define state $\Psi(x) = [\psi(x) , (1/\kappa_0) \cdot d\psi/dx]^T$ with $\kappa_0 = \omega / c$. The transfer matrix across one barrier (B) and one well (W) are:
- Barrier: $M_B = [[\cosh(\kappa d), (1/\kappa) \sinh(\kappa d)], [\kappa \sinh(\kappa d), \cosh(\kappa d)]]$
- Well: $M_W = [[\cos(k s), (1/k) \sin(k s)], [-k \sin(k s), \cos(k s)]]$
- with $k = \omega / c$ in wells and $\kappa^2 = (\omega_0^2 - \omega^2) / c^2$ in barriers; $\omega_0 = \mathrm{sqrt}(2\ k_{const})$ from the NLEP mass gap

(k_const is the NLEP coefficient).

3. Bloch Condition and Miniband Formation

- For an infinite superlattice, Bloch's theorem gives det| M_cell − e^(i q a) I | = 0 with M_cell = M_B · M_W. The dispersion relation becomes:
- cos(q a) = cosh(κ d) · cos(k s) + ((κ^2 + k^2)/(2 κ k)) · sinh(κ d) · sin(k s)
- Allowed minibands satisfy |cos(q a)| ≤ 1; band gaps occur when |cos(q a)| > 1.

4. Finite N: Multi-Barrier Resonant Tunneling

- For N cells between leads, the total transfer matrix is M_N = (M_cell)^N. With incidence from the left, the coherent transmission amplitude is T_amp = 1 / |(M_N)11|, and the transmission probability T = |T_amp|^2. Sharp resonances occur when the well phase satisfies k s = m π (m ∈ ℤ) and κ d is moderate, yielding miniband-assisted tunneling.

5. AC Pressure Bias and Dynamic Localization

- Apply an oscillatory pressure bias ΔP(t) = ΔP0 cos(Ω t) across the lattice. The per-cell phase obeys d(Δθ)/dt = ΔP(t)/ρ_eff, so the well phase shifts k s → k s + β sin(Ω t) with β = ΔP0 / (ρ_eff Ω). Time-averaged transmission becomes a Bessel-weighted sideband sum (dynamic localization / photon-assisted tunneling analogue).

6. Nonlinear Undercut and Self-Induced Transparency

- A co-propagating support displacement ε_sup lowers effective barrier pressure: P_eff = k · (ε_bar + ε_sup)^2. If 2 ε_bar ε_sup + ε_sup^2 < 0, the decay constant κ decreases and transparency windows open inside gaps (nonlinear resonant tunneling).

7. Losses and Resonance Q-Factor

- Small viscous loss γ_AM broadens resonances. The FWHM scales approximately as Γ ≈ 2 γ_AM · (∂ω/∂q) / a. Ultra-high K_AM and tiny γ_AM produce high-Q resonances and long coherence lengths.

8. Design Relations (Equation-First)

- • First miniband center: ω_c ≈ sqrt(ω0^2 + (π c / s)^2) for κ d ≈ 1.
- • Finite-stack resonance: k s ≈ m π and sinh(κ d) ≲ (2 k κ)/(κ^2 + k^2).
- • Trends: bandgap width ↑ with κ d and barrier fraction d/a; miniband width ↑ with s and ↓ with κ.

9. Physical Summary

- AM superlattices—periodic pressure-barrier structures—create controllable minibands and gaps for AM displacement modes. Multi-barrier resonant tunneling follows from coherent phase matching, and can be tuned by AC pressure bias and nonlinear undercut. The ultra-stiff, low-loss

Medium supports sharp, high-Q transport windows.

Photon Coherence & Mode Stability

1. Photon as a Transverse AM Displacement Mode

In the AM Model, a photon is a stable transverse ripple of the Absolute Medium. Its displacement field is:

- $\varepsilon_{ph}(x,t) = A \cos(k \cdot x - \omega t)$
- with $\omega = c\,k$, where $c = \sqrt{K_{AM} / \rho_{AM}}$.

2. Coherence Condition: Phase Preservation

Photon coherence requires phase stability across propagation:

- $\Delta\phi = k\,\Delta x - \omega\,\Delta t = \text{constant}$
- The AM stiffness K_{AM} ensures negligible phase drift.

3. Coherence Length from AM Parameters

Define coherence length L_{coh} by phase spread ≤ 1 rad:

- $L_{coh} \approx c\,\tau_{coh}$

AM damping is extremely small, so:

- $\tau_{coh} \propto 1 / \gamma_{AM}$
- where $\gamma_{AM} \ll 1$ due to near-zero viscous loss.

4. Mode Stability Equation (NLEP Background)

Photon modes propagate in a weakly nonlinear Medium governed by:

- $\partial^2\varepsilon/\partial t^2 - c^2\,\nabla^2\varepsilon + 2k\,\varepsilon = 0$
- The nonlinear term $2k\varepsilon$ produces self-correction, stabilizing amplitude for small ε.

5. Suppression of Cross-Mode Scattering

Overlap of two photon modes ε_1, ε_2 yields cross pressure:

- $P_{cross} = 2k\,\varepsilon_1\,\varepsilon_2$
- Coherent photons satisfy:
- $\int P_{cross}\,dV \to 0$
- ensuring no mode mixing and maintaining coherence.

6. Stability Against Grain-Scale Disturbances

Let $\eta(x)$ be grain-scale AM fluctuation. Photon stability requires:

- $|\eta| \ll |\varepsilon_{ph}|$
- Pressure curvature adds restoring force:
- $F_{rest} = -\nabla(k\,\varepsilon_{ph}^2)$

7. AM Coherence Criterion

Combining the above, coherence holds when:

- $(\gamma_{AM} / K_{AM}) \ll (A\,k)$
- ensuring phase stability and mode integrity over large distances.

8. Physical Summary

Photon coherence in the AM Model arises from extreme Medium stiffness, negligible damping, and weak nonlinear elastic corrections. These combine to create long-range stability of transverse modes, preserving both amplitude and phase across cosmological distances.

Quantum Entanglement as Nonlocal Pressure-Field Coupling

1. Entanglement as Shared AM Displacement Field

In the AM Model, two particles become entangled when they share a non-separable displacement field of the Absolute Medium (AM):

- $\varepsilon_{total}(x,t) = \varepsilon_A(x,t) + \varepsilon_B(x,t) + \varepsilon_{AB}(x,t)$
- where ε_{AB} is the coupled, non-factorizable cross-field that links A and B.

2. Nonlocal Pressure Coupling Term

The AM pressure field is nonlinear under NLEP:

- $P = k\,\varepsilon^2$
- Thus the combined field produces cross-pressure:
- $P_{cross} = 2k\,\varepsilon_A\,\varepsilon_B$
- This term persists even when A and B are spatially separated, provided the AM displacement kernel remains coherent.

3. Entanglement Condition (Non-Separable Field)

Entanglement occurs when the total field cannot be written as a product of individual fields:

- $\varepsilon\_total \neq f\_A(\varepsilon\_A) \cdot f\_B(\varepsilon\_B)$
- Mathematically, this corresponds to non-vanishing second functional derivative:
- $\delta^2 P\_total / (\delta\varepsilon\_A \, \delta\varepsilon\_B) = 2k \neq 0$

4. Nonlocal Correlation Equation

Pressure-curvature quantum potential for the coupled system is:

- $Q\_AM = -(1/\rho\_AM) \nabla^2(k \, \varepsilon\_total^2)$
- Expanding, the AB correlation term is:
- $Q\_AB = -(2k / \rho\_AM) \nabla^2(\varepsilon\_A \, \varepsilon\_B)$
- This term propagates through the Medium at stiffness-determined speed c.

5. Measurement and Mode Lock-In

Measurement forces displacement into the Saturation Layer ($\varepsilon \rightarrow \varepsilon\_sat$):

- $\sigma = k \, \varepsilon\_sat^2$
- The nonlinear collapse condition is:
- $\partial P\_total/\partial\varepsilon\_A \rightarrow$ large $\Rightarrow \varepsilon\_B$ forced into correlated mode

6. Entanglement Decay (Decoherence)

Decoherence corresponds to decay of ε_AB due to AM damping γ_AM:

- $d\varepsilon_{AB}/dt = -\gamma_{AM}\,\varepsilon_{AB}$
- Long-range entanglement corresponds to $\gamma_{AM} \ll 1$, due to near-zero viscous loss in AM.

7. Physical Summary

In the AM Model, entanglement emerges from a shared nonlocal displacement field producing cross-pressure terms that link particle states. Measurement collapse arises from nonlinear saturation forcing mode alignment. Decoherence reflects damping of the shared AM mode. This provides a mechanical interpretation of quantum entanglement rooted in the Medium's coupling laws.

Relativistic AM Dispersion Relations (Photon & Massive Modes)

1. AM Displacement Field ε(x,t)

All excitations in the Absolute Medium (AM) arise from the displacement field:

- $\varepsilon = \varepsilon(x,t)$
- relating directly to wave propagation, mass generation, and relativistic dynamics.

2. Fundamental Relativistic Wave Equation

The AM hyperbolic field equation governing all modes is:

- $\partial^2\varepsilon/\partial t^2 - c^2\,\nabla^2\varepsilon + 2k\,\varepsilon = C_v\,(\Delta\rho)$
- with $c = \sqrt{(K_{AM} / \rho_{AM})}$.

For free excitations ($\Delta\rho = 0$):

- $\partial^2\varepsilon/\partial t^2 - c^2 \nabla^2\varepsilon + 2k\,\varepsilon = 0$

3. Photon Dispersion Relation (Transverse Mode)

For small-amplitude transverse waves ($\varepsilon \ll \varepsilon_{sat}$), the nonlinear term is negligible:

- $\partial^2\varepsilon/\partial t^2 - c^2 \nabla^2\varepsilon = 0$
- Insert plane-wave solution $\varepsilon = A\, e^{i(k\cdot x - \omega t)}$:
- $\omega^2 = c^2 k^2$
- This is the relativistic photon dispersion relation in AM.

4. Massive AM Modes (Longitudinal / Confined)

For confined or quasi-longitudinal modes, the nonlinear elasticity (NLEP) contributes a mass-generating term:

- $\partial^2\varepsilon/\partial t^2 - c^2 \nabla^2\varepsilon + 2k\,\varepsilon = 0$
- Assume $\varepsilon = A\, e^{i(k\cdot x - \omega t)}$:
- $\omega^2 = c^2 k^2 + 2k$
- Define effective mass m_{eff} via:
- $(m_{eff}\, c^2 / \hbar)^2 = 2k$
- Thus massive mode dispersion becomes:
- $\omega^2 = c^2 k^2 + (m_{eff}\, c^2 / \hbar)^2$

5. Unified AM Dispersion Relation

Both photon and massive excitations satisfy a unified relation:

- $\omega^2 = c^2 k^2 + \omega_0^2$
- with $\omega_0 = 0$ for photon modes,

- $\omega_0 = \sqrt{(2k)}$ for massive modes.

6. Group Velocity and Relativistic Limits

Group velocity is:

- $v_g = d\omega/dk = (c^2 k) / \omega$

For photon modes ($\omega_0 = 0$):

- $v_g = c$

For massive modes:

- $v_g = c \sqrt{(1 - (\omega_0^2 / \omega^2))}$
- replicating the relativistic energy–momentum relation.

7. Energy–Momentum Relation in AM

Identify ω with energy $E/\hbar$ and k with momentum $p/\hbar$:

- $E^2 = p^2 c^2 + m_{eff}^2 c^4$
- Thus AM massive modes obey the standard relativistic dispersion curve.

8. Physical Summary

Photon modes propagate as massless transverse AM waves with $\omega = ck$, while massive modes arise from nonlinear elasticity and obey $\omega^2 = c^2k^2 + \omega_0^2$. This unified dispersion framework reproduces standard relativistic physics while grounding it in mechanical properties of the Absolute Medium.

AM Waveguides & Bound Photonic States in Matter Structures

1. Photon as AM Transverse Displacement Mode

In the AM Model, a photon is a transverse displacement ripple in the Absolute Medium (AM):

- $\varepsilon\_ph(x,t) = A \cos(k \cdot x - \omega t)$
- with dispersion $\omega = c\, k$, where $c = \sqrt{(K\_AM / \rho\_AM)}$.

2. AM Waveguide: Matter-Induced Boundary Condition

Matter creates local AM stiffness contrast $\Delta K(x)$. A waveguide is a region where:

- $K\_eff(x) = K\_AM + \Delta K(x)$
- and $\partial K\_eff/\partial\perp$ produces confinement via pressure curvature.
- The confinement condition is:
- $\nabla\perp^2 \varepsilon + (\omega^2/c^2)\, \varepsilon < 0$
- which forces the mode to remain in the high-K_AM channel.

3. AM Pressure Curvature as Guiding Force

NLEP pressure is $P = k\, \varepsilon^2$. Waveguide confinement arises from curvature:

- $F\_guide = -\nabla\perp P = -2k\, \varepsilon\, \nabla\perp\varepsilon$
- This pulls the photon inward toward the waveguide core.

4. Bound State Condition in AM Waveguides

A bound photonic state is a standing AM mode satisfying:

- $\nabla^2 \varepsilon + (\omega^2/c^2)\, \varepsilon = 0$ inside core
- $\nabla^2 \varepsilon + (\omega^2/c\_eff^2)\, \varepsilon = 0$ outside core
- where c_eff < c due to ΔK and Δρ of matter.

5. Quantized Bound Modes

Boundary continuity yields quantization conditions:

- $k\_n\, R = n\, \pi - \phi$
- with ϕ determined by stiffness contrast and AM impedance mismatch.

6. Energy of Bound Photonic States

Mode energy is given by the AM dispersion relation:

- $E\_n = \hbar\, \omega\_n = \hbar\, c\, k\_n$
- Bound states correspond to discrete k_n allowed by the waveguide geometry.

7. Loss Minimization from AM Stiffness

Lossless transmission occurs when AM viscous coefficient $\gamma\_AM \ll k\_n\, c$:

- $Q\_factor \approx \omega\_n / (2\, \gamma\_AM) \rightarrow$ extremely large
- giving near-perfect coherence along matter-defined channels.

8. Physical Summary

Matter structures alter local AM stiffness and density, forming natural waveguides. Photons propagate as confined transverse displacement modes whose quantization arises from AM boundary constraints. Ultra-high stiffness and minimal damping make AM waveguides capable of supporting long-range, coherent bound photonic states.

Zero-Point Fluctuations as Grain-Scale Breathing Modes

1. Fundamental Grain-Scale Structure of the AM

The Absolute Medium (AM) possesses a microscopic grain scale L_{fund}. Each grain supports intrinsic oscillations of the displacement field $\varepsilon(x,t)$. The natural frequency of a grain is:

- $\omega_{fund} = \pi c / L_{fund}$
- where $c = \mathrm{sqrt}(K_{AM} / \rho_{AM})$ is the AM wave speed.

2. Definition of Zero-Point Breathing Mode

A single AM grain supports a minimal ("zero-point") breathing oscillation even in the absence of external excitation:

- $\varepsilon_{fund}(t) = \varepsilon_0 \cos(\omega_{fund}\, t)$
- The corresponding NLEP pressure is:
- $P_{fund} = k\, \varepsilon_0^2$

3. Zero-Point Energy Density from Fundamental Mode

For each grain with volume V_fund ≈ L_fund^3, the zero-point energy is:

- E_0 = (1/2) ħ ω_fund
- Thus, the zero-point energy density is:
- ρ_0 = E_0 / V_fund = (ħ ω_fund) / (2 L_fund^3)

4. Pressure Curvature from Breathing Modes

Curvature of pressure from grain oscillation is:

- ∇^2P_fund = 2k (ε_fund ∇^2ε_fund + |∇ε_fund|^2)
- At grain scale, ∇^2ε_fund ~ ε_0 / L_fund^2.

5. Background Pressure as Zero-Point Sum

Summing over all grains gives the baseline AM pressure:

- P_Λ = ξ ρ_0 = ξ (ħ ω_fund) / (2 L_fund^3)
- where ξ ≈ 1 encodes the Medium's ultra-stiff equation of state.

6. Influence on AM Wave Propagation

Zero-point breathing causes small-scale fluctuations superimposed on long-wavelength AM waves. Effective stiffness becomes:

- K_eff = K_AM + δK_ZP
- with correction term:
- δK_ZP ∝ k ε_0^2 / L_fund^2

7. Zero-Point Modes Enable Undercut Processes

Breathing fluctuations $\eta(x) = \varepsilon\_0 \cos(\omega\_\text{fund}\ t)$ enter tunneling and collapse equations via:

- $\varepsilon(x) = \varepsilon\_\text{classical}(x) + \eta(x)$
- and modify local pressure:
- $P = k\ (\varepsilon\_\text{classical} + \eta)^\wedge 2$.

8. Summary

Zero-point fluctuations in the AM Model correspond to intrinsic grain-scale breathing modes with frequency $\omega\_\text{fund} = \pi c/L\_\text{fund}$. These fluctuations generate finite zero-point pressure and energy density, contribute to AM stiffness renormalization, and act as the mechanical origin of quantum phenomena such as tunneling and mode selection.

Quantum Vacuum Drag & AM Grain Interference

1. Grain Structure of the Absolute Medium (AM)

The AM consists of discrete stiffness grains of characteristic size L_fund. Each grain supports intrinsic zero-point breathing modes:

- $\varepsilon\_\text{fund}(t) = \varepsilon\_0 \cos(\omega\_\text{fund}\ t)$
- $\omega\_\text{fund} = \pi\ c\ /\ L\_\text{fund}$

2. Zero-Point Pressure Field P_ZP

The zero-point NLEP pressure inside each grain is:

- $P\_ZP = k\,\varepsilon\_0^2$
- Spatial fluctuations arise from grain-to-grain variation of ε_0.

3. Pressure Gradient Noise from Grain Interference

When a particle moves through the AM lattice, its displacement $\varepsilon\_part(x)$ interacts with grain-scale fluctuations $\eta(x)$:

- $\varepsilon\_total(x) = \varepsilon\_part(x) + \eta(x)$

The resulting total pressure is:

- $P(x) = k\,(\varepsilon\_part + \eta)^2$

Expanding yields an interference term:

- $P\_int = 2k\,\varepsilon\_part\,\eta$

4. Drag Force from Zero-Point Pressure Curvature

The effective drag force is the spatial gradient of the interference pressure:

- $F\_drag = -\nabla P\_int = -2k\,\nabla(\varepsilon\_part\,\eta)$
- Averaging over grain fluctuations η gives a non-zero net drag.

5. Velocity Scaling of AM Vacuum Drag

For a particle moving with velocity v through grain oscillations:

- $\eta(x,t) = \varepsilon\_0 \cos(\omega\_fund\ t - k\_fund\ x)$

The drag becomes:

- $F\_drag \propto - k\ \varepsilon\_0\ v / L\_fund$
- This term is extremely small due to tiny ε_0 and L_fund.

6. Interference Pattern Between Grains

Adjacent grain breathing modes superpose:

- $\eta\_total = \Sigma\_j\ \varepsilon\_0 \cos(\omega\_fund\ t + \phi\_j)$

giving a fluctuating pressure landscape:

- $P\_ZP(x) = k\ \eta\_total^2$
- Particles experience micro-scale jitter analogous to vacuum jitter (zitterbewegung).

7. Stochastic AM Equation of Motion

The particle's motion obeys a Langevin-type equation:

- $m\ d^2x/dt^2 = F\_classical + F\_drag + F\_noise$

with:

- $F\_noise = - 2k\ \nabla(\varepsilon\_part\ \eta)$

8. Physical Summary

Quantum vacuum drag in the AM Model arises from interference between particle displacement fields and grain-scale zero-point breathing oscillations. This produces tiny fluctuating pressure gradients that generate stochastic drag and jitter, offering a mechanical explanation for vacuum friction and quantum randomness.

AM Phonon–Photon Coupling in Dense Matter

1. AM Field Components in Dense Matter

In the AM Model, both photons and phonons arise from the displacement field ε(x,t) of the Absolute Medium (AM). In dense matter, ε splits into:

- ε(x,t) = ε_photon(x,t) + ε_phonon(x,t) + ε_int(x,t)
- where ε_int represents the interaction (coupling) displacement.

2. Photon Mode (Transverse AM Ripple)

Free photon in AM is given by:

- ε_photon = A cos(k·x − ωt)

with dispersion:

- $\omega^2 = c^2 k^2$

3. Phonon Mode (Longitudinal Matter-Coupled Mode)

In dense matter, local compression produces longitudinal AM–matter vibrations:

- $\varepsilon\_phonon = B \cos(q \cdot x - \Omega t)$

with dispersion:

- $\Omega^2 = c\_s^2 q^2$
- where $c\_s < c$ is the AM–matter sound speed.

4. Nonlinear Coupling via Pressure Interaction

NLEP pressure: $P = k \varepsilon^2$. Combined fields produce cross-pressure:

- $P\_int = 2k\ \varepsilon\_photon\ \varepsilon\_phonon$
- This term mediates photon–phonon energy exchange.

5. Coupling Equation (Mode Mixing)

Full AM field equation in matter:

- $\partial^2\varepsilon/\partial t^2 - c^2 \nabla^2\varepsilon + 2k\,\varepsilon = C\_v\,(\Delta\rho)$

Insert $\varepsilon = \varepsilon\_photon + \varepsilon\_phonon$ to obtain mixing terms:

- $\partial^2\varepsilon\_photon/\partial t^2 - c^2 \nabla^2\varepsilon\_photon = -\ 2k\ \varepsilon\_phonon\ \varepsilon\_photon$
- $\partial^2\varepsilon\_phonon/\partial t^2 - c\_s^2 \nabla^2\varepsilon\_phonon = -\ 2k\ \varepsilon\_photon\ \varepsilon\_phonon$

6. Energy Exchange Rate

Define interaction Hamiltonian density:

- $H\_int = k\,\varepsilon\_photon^2\,\varepsilon\_phonon^2$

Energy transfer rate:

- $dE/dt \propto \int \varepsilon\_photon\,\varepsilon\_phonon\,dV$

7. Bound Photon–Phonon Hybrid (AM Polaritonic State)

Stable hybrid mode occurs when:

- $\omega \approx \Omega \Rightarrow$ phase-locking condition

Hybrid displacement:

- $\varepsilon\_hybrid = \alpha\,\varepsilon\_photon + \beta\,\varepsilon\_phonon$

with eigenfrequency:

- $\omega\_h^2 = (\omega^2 + \Omega^2)/2 \pm \mathrm{sqrt}[(\omega^2 - \Omega^2)^2/4 + g^2]$
- where coupling strength $g \propto k \int \varepsilon\_photon\,\varepsilon\_phonon\,dV$.

8. Physical Summary

Phonon–photon coupling in the AM Model arises naturally from nonlinear pressure interactions. Dense matter modifies AM stiffness, allowing transverse (photon) and longitudinal (phonon) displacement modes to exchange energy and form hybrid bound states. This reproduces polariton-like behavior using purely mechanical Medium interactions.

AM Josephson-Like Effects in Coherent Matter Flows

1. Coherent AM–Matter Phase Field

In the AM Model, coherent matter flow is described by a phase-bearing displacement field:

- $\varepsilon(x,t) = A(x) \cos(\theta(x,t))$
- The phase $\theta(x,t)$ plays the same mathematical role as the quantum phase in superconducting systems.

2. AM Phase Gradient → Flow Velocity

Flow of coherent matter corresponds to a phase gradient in ε:

- $v = (1/\rho\_eff)\ \nabla\theta$
- where ρ_eff is the effective mass–coupled AM density.

3. Two-Region Coupling: AM Josephson Geometry

Consider two coherent AM–matter regions separated by a thin barrier of width d:

- Region L: $\varepsilon\_L = A\_L \cos(\theta\_L)$
- Region R: $\varepsilon\_R = A\_R \cos(\theta\_R)$
- Their overlap generates an interference pressure:
- $P\_int = 2k\ \varepsilon\_L\ \varepsilon\_R = 2k\ A\_L\ A\_R \cos(\theta\_L - \theta\_R)$

4. AM Josephson Current (Pressure-Driven Flow)

The AM analogue of Josephson current is obtained from the gradient of interference pressure:

- $J\_AM = -\nabla P\_int$
- For a 1-D weak link:
- $J\_AM = J\_0 \sin(\Delta\theta)$
- with $\Delta\theta = \theta\_L - \theta\_R$ and $J\_0 \propto k\ A\_L\ A\_R$.

5. Phase Evolution and AC Josephson Effect

A pressure imbalance ΔP across the barrier induces phase evolution:

- $d(\Delta\theta)/dt = \Delta P / \rho\_eff$
- Thus the AM current oscillates as:
- $J\_AM(t) = J\_0 \sin(\Delta\theta\_0 + (\Delta P/\rho\_eff)t)$

6. Nonlinear NLEP Reinforcement

NLEP pressure $P = k\,\varepsilon^2$ reinforces phase locking:

- $d^2\theta/dt^2 + \omega\_J^2 \sin(\Delta\theta) = 0$
- with Josephson-like frequency $\omega\_J^2 = (2k\ A\_L\ A\_R)/\rho\_eff$.

7. Quantum-Classical Crossover via Saturation Layer

When displacement approaches saturation $\varepsilon \rightarrow \varepsilon\_sat$, collapse eliminates multi-phase superposition:

- $\Delta\theta \rightarrow 0$ or π
- yielding classical AM phase locking analogous to the DC

Josephson state.

8. Physical Summary

Josephson-like effects in the AM Model arise from nonlinear interference pressure between coherent matter flows. Phase gradients drive AM current, pressure bias produces oscillatory (AC) Josephson behavior, and NLEP locks phases into stable DC-like states. The phenomenon emerges mechanically from AM displacement coupling rather than quantum postulates.

AM Probability Current from ε-Mode Flow

1. Displacement Field as Fundamental AM State Variable

The AM Model uses the Medium displacement field $\varepsilon(x,t)$ as the fundamental descriptor of both photonic and matter-wave behavior:

- $\varepsilon = \varepsilon(x,t)$
- This field governs pressure, flow, and coherent transport through the AM.

2. AM Wave Equation (Free Field)

Free propagation of ε obeys the linearized AM wave equation:

- $\partial^2\varepsilon/\partial t^2 - c^2 \nabla^2\varepsilon = 0$
- with c = sqrt(K_AM / ρ_AM).

3. Complex Mode Representation

Introduce a complex representation to track amplitude and phase:

- $\psi(x,t) = \varepsilon(x,t)\, e^{i\theta(x,t)}$
- The physical field is obtained as the real part, but ψ simplifies transport analysis.

4. AM Continuity Equation

Start from energy density of the AM ripple:

- $\rho_E = \frac{1}{2} (\rho_{AM} |\partial\varepsilon/\partial t|^2 + K_{AM} |\nabla\varepsilon|^2)$

Using the wave equation and differentiating w.r.t time yields:

- $\partial\rho_E/\partial t + \nabla \cdot J_{AM} = 0$
- which is the continuity equation for AM wave energy.

5. AM Probability Current Definition

Define ε as the amplitude, θ as the phase. Using $\psi = \varepsilon\, e^{i\theta}$, compute:

- $J_{AM} = (K_{AM} / \rho_{AM}) \cdot \mathrm{Im}(\psi^* \nabla\psi)$

Expanding gives:

- $J_{AM} = (K_{AM} / \rho_{AM})\, \varepsilon^2 \nabla\theta$
- Thus AM probability current is proportional to phase gradient of the displacement field.

6. Velocity Interpretation

Define effective flow velocity:

- $v\_AM = \nabla\theta$

Then the current becomes:

- $J\_AM = (K\_AM / \rho\_AM)\, \varepsilon^2\, v\_AM$

7. Relation to Bohmian Current

The Bohmian quantum current is:

- $J\_B = (\hbar/m)\, |\psi|^2\, \nabla\theta$

AM current reproduces the same structure via the Medium parameters:

- $(\hbar/m) \leftrightarrow (K\_AM / \rho\_AM)$

8. Coherent Flow Through Barriers

Inside tunneling regions, ε decreases but phase gradient increases, keeping J_AM approximately constant:

- $J\_AM \approx$ const.
- Matching boundary conditions ensures flux conservation.

9. Physical Summary

AM probability current emerges from the phase-gradient–driven transport of displacement energy. The continuity law follows from the AM wave equation, and the current expression matches the structure of quantum mechanical probability flow, but is derived mechanically from Medium dynamics.

Turning AM into Testable Science: Predictions vs GR/QFT/ΛCDM and Engagement with Current Constraints

Purpose & Scope

Address the recommendations to: (i) derive falsifiable predictions that differ from GR, QFT, and ΛCDM; and (ii) engage existing constraints (equivalence principle, light-speed isotropy, and GW propagation).

Minimal Parameterization Used Here

Core relations: Dissociative Pressure $P_D = k (\Delta\rho)^2$; Non-Linear Elasticity (NLEP) $\sigma = k\ \varepsilon^2$; Structural Density Law (SDL) $\rho_{AM} \propto M^2/R^5$; Volumetric Coupling $C_v = \kappa/L_{fund}$; relativized AM wave equation for ε.

A. Testable Predictions that Differ from GR / QFT / ΛCDM

A1. Second-Order Light-Speed Anisotropy from AM Density Gradients

In AM, $c_{eff}^2 = K_{AM} / \rho_{AM}$. SDL implies tiny spatial gradients in ρ_{AM} at heliocentric scales, giving direction-dependent c_{eff}.

$c_{eff}(\hat{n}) \approx c \cdot [\ 1 - (1/2) \cdot \delta_{rho}(\hat{n})\ ]$ (1)

A2. Weak-Equivalence-Principle (WEP) Shift via Volumetric Coupling

AM predicts composition/structure dependence through $C_v(\rho_m)$:

$\eta\_AM \sim |\Delta f(\rho\_m)| \cdot [\ P\_D / (\rho\_AM \cdot g)\ ]$ (2)

A3. GW Phase–Amplitude Correlation from NLEP

Nonlinear stiffness induces weak dispersion for finite amplitude:

$\omega\text{^}2 = c\text{^}2\ k\text{^}2 + 2\ k\_NLEP \cdot \langle \varepsilon\text{^}2 \rangle$ (3)

A4. Rotation-Curve Residual Pattern (SDL + VCP)

AM predicts residuals correlated with $d[\ C\_v(r)\ \rho\_AM(r)\ ]/d \ln r$ rather than stochastic ΛCDM halo substructure.

A5. UV Cutoff Signature Near ω_max

$\omega\_max = \pi\ c / L\_fund$ (4)

Look for turnover/plateau in vacuum-noise slope approaching ω_max (lab lower bounds; astro preferred for reach).

B. Engagement with Existing Constraints (Design Principles)

WEP tests (torsion/satellite), light-speed isotropy (cavity), GW propagation (LIGO/Virgo/KAGRA; LISA; PTAs), cosmology (rotation curves, lensing, clusters).

C. Experiment & Analysis Protocols

C1. Differential WEP Test Maximizing Δf(ρ_m)

$\eta(h) \propto \Delta f \cdot [\, P\_D(h) / (\rho\_AM \cdot g(h)) \,]$ (5)

C2. Targeted Light-Speed Isotropy Demodulation

Demodulate cavity $\delta\nu/\nu$ at sidereal/annual frequencies with SDL templates.

C3. GW Phase–Amplitude Regression

$\partial\phi\_res/\partial E\_strain \neq 0$ (AM); $= 0$ (GR) (6)

C4. Broadband Optomechanical Noise Survey

$L\_fund = \pi c / \omega\_max$ (7)

1. Order-of-Magnitude Estimates (Signal Sizes and Detectability)

These OOM estimates use plausible parameter ranges and indicate current/near-future detectability. They will be narrowed after calibration of C_SDL and bounds on k_NLEP.

1.1 A1 — Light-Speed Anisotropy

$\delta c/c \approx 0.5 \cdot \delta\_rho$ (8)

Assume δ_rho at 1 AU ~ 1e-20 to 1e-16 (pre-calibration). Then δc/c ~ 5e-21 to 5e-17. Current cryogenic cavities approach ~1e-18 fractional sensitivity; detection possible on upper range with template-locked ephemeris demodulation.

1.2 A2 — WEP Shift

Define $Xi := P_D/(\rho_{AM}\, g)$. Using $f(\rho_m)$ from Section 4, $|\Delta f|$ ranges from ~0.03 (dense vs water) up to ~80 (aerogel vs tungsten). With $Xi \sim 1e\text{-}15$ to $1e\text{-}12$ depending on geometry, predict $\eta_{AM} \sim 2e\text{-}17$ to $8e\text{-}11$. Near-term torsion sensitivity ~1e-13 suggests discovery space for large $|\Delta f|$ and enhanced Xi (e.g., near mass anomalies).

1.3 A3 — GW Phase–Amplitude Correlation

Representative chirp: $f \sim 100$ Hz ($\omega \sim 600\ s^{-1}$), duration $T \sim 0.2$ s, strain $h \sim 1e\text{-}21$. With $\langle \varepsilon^2 \rangle \sim h^2$, additional phase scales as $\Delta\phi \sim (k_{NLEP}\, h^2 / \omega)\, T$. A stacked bound $|\Delta\phi| < 0.01$ rad yields $k_{NLEP} < 3e43$ (native units). Section 3 converts to a dimensionless bound $\tilde{k} = k_{NLEP}/(\rho_{AM}\, c^2)$.

1.4 A4 — Rotation-Curve Residuals

Assuming gentle radial gradients in C_v and ρ_{AM} consistent with SDL halos gives $\Delta v/v \sim 1e\text{-}2$ to $1e\text{-}1$ in outer disks; testable via residual maps correlated with baryonic geometry (contrasts with ΛCDM stochastic substructure).

1.5 A5 — UV-Cutoff Turnover

$$\omega_{max} = \pi c / L_{fund} \quad (9)$$

If lab probes reach $\omega_{probe} \sim 1e12\text{–}1e14\ s^{-1}$ without turnover, then $L_{fund} > \pi c / \omega_{probe} \sim 1e\text{-}6\text{–}1e\text{-}8$ m (weak bound). Methodology stands; astrophotonic frequencies provide stronger reach.

2. Calibrate C_SDL from Solar-System Data

Calibrate SDL normalization via perihelion precession or Shapiro delay to fix ρ_AM at AU scales, turning A1 into a quantitative prediction.

Perihelion-precession route (outline):

- Use ρ_AM(R) = C_SDL · M_sun^2 / R^5 and P_D(R) ∝ M_sun^2 / R^4.

- Derive a_AM(R) from pressure gradient and coupling; map to secular Δvarpi(C_SDL; a, e) via Lagrange planetary equations.

- Fit to residual precession to solve/bound C_SDL.

Shapiro-delay route (outline):

- Define n_eff ≈ 1 + A · ρ_AM.

- Integrate Δt_AM ≈ ∫(n_eff − 1) ds with ρ_AM from SDL.

- Compare with ranging residuals to solve for C_SDL.

With C_SDL fixed, compute δ_rho(1 AU) and hence δc/c via Eq. (1); implement ephemeris-locked anisotropy templates for cavity tests.

3. Bound k_NLEP from Existing GW Data

- Construct GR-subtracted residual phase ϕ_res(f) from public strain.

- Regress ϕ_res against E_strain ∝ h(f)^2 across time-frequency pixels and events.

- If slope is null within σ_ϕ, set k_NLEP < $(\sigma\_\phi \cdot \omega)/(h^2 \cdot T)$ for each event; combine for a global bound.

- Report also k_tilde = k_NLEP/(ρ_AM c^2) using manuscript ρ_AM.

4. Material Contrast Quantification (Converts η(h) to a Number)

Use f(ρ_m) = 1 + (ρ_m/ρ_ref)^n with n in (1,2) and ρ_ref ~ 1e3 kg/m^3.

- Aerogel (ρ ~ 1e2 kg/m^3): f ≈ 1.03

- Aluminum (ρ ~ 2.7e3 kg/m^3): f ≈ 5.44

- Tungsten (ρ ~ 1.93e4 kg/m^3): f ≈ 85.8

Max contrast example: Δf (aerogel vs tungsten) ≈ 84.76 for n=1.5, ρ_ref=1000 kg/m^3.

Definite site prediction: define Xi(h) = P_D(h)/(ρ_AM g(h)) from local AM-field modeling near a mass anomaly; then η(h) = Δf · Xi(h). Provide η(h) curve and target sensitivity for the chosen facility.

AM Model — Symbols List

1. Fields and State Variables

- **ε(x,t)** — AM displacement field [dimensionless]
- **ψ = ε e^{iθ}** — Complex amplitude [dimensionless]
- **θ(x,t)** — Phase [radians]
- **ρ_E** — Energy density [J/m^3]

2. Medium Properties (Fundamental)

- **ρ_{{AM},0}** — Background AM density [kg/m^3] $\approx 1.2\times10^{17}$
- **κ_{AM}** — AM stiffness = ρ_{{AM},0} c^2 [J/m^3] $\approx 1.1\times10^{43}$
- **c** — Wave speed = √(κ_{{AM}}/ρ_{{AM},0}) [m/s] $\approx 3\times10^{8}$
- **k_{NLEP}** — NLEP elasticity coefficient [J/m^3] $\approx 10^{40}$ (α_{{NLEP}}·κ_{{AM}})

3. Derived / Effective Parameters

- **C_v** — Volumetric coupling constant [m^{-1}] $\approx 10^{-7}$
- **C_{SDL}** — SDL constant [m^2/kg] (derived from fits)
- **α** — Kernel amplitude factor [m^2] (from $\varepsilon \approx \alpha/r^2$)
- **α_{NLEP}** — Nonlinearity parameter [dimensionless] $\approx 10^{-3}$
- **ν** — Coupling function [dimensionless]

4. Flow Variables

- **v_r, v_θ** — Radial/tangential flow speeds [m/s]
- **Q** — Sink strength [m^3/s]
- **Γ** — Circulation [m^2/s]

5. Wave and Dispersion

- **ω** — Angular frequency [rad/s]
- **k** — Wavenumber [m^{-1}] (distinct from k_{{NLEP}})
- **κ** — Evanescent decay constant [m^{-1}]
- **q** — Bloch wavenumber [m^{-1}]

6. Thermodynamic / Cosmological

- **P_Λ** — Residual background pressure [Pa] ≈ 10^{-9}
- **Λ_{AM}** — AM cosmological constant [m^{-2}] = (8π G / c^4) P_Λ
- **R(t)** — Scale factor [dimensionless]
- **T_{CMB}** — CMB temperature [K]

7. Tensor Components

- **T^{AM}_{mu u}** — Full AM stress–energy tensor [J/m^3]
- **E_{mu u}** — Elastic sector [J/m^3]
- **F_{mu u}** — Flow sector [J/m^3]
- **S_{mu u}** — Density–gradient sector [J/m^3]
- **C_{mu u}** — Coupling sector [J/m^3]

8. Quantum Analogues

- **Q_{AM}^{(acc)}** — Acceleration-like quantum potential [s^{-2}]
- **Q_{AM}^{(E)}** — Energy-density quantum potential [J/m^3]
- **P_{cross}** — Interference pressure [J/m^3]
- **Q_{AB}** — Correlation curvature [J/m^3]
- **J_{AM}** — Probability/energy current [J/(m^2·s)]

9. Coupled–Mode Parameters

- **Ω** — Phonon mode frequency [rad/s]
- **g** — Mode coupling strength [rad/s]
- **J_0** — Josephson critical current [kg/(m·s^2)?]
- **Δθ** — Phase difference [rad]
- **ρ_{eff}** — Effective coupled density [kg/m^3]

- **A_L, A_R** — Left/right coherent amplitudes (Josephson-like weak link)

10. Jet / Unified-Field / Fundamental Length Additions

- **A_{jet}** — Jet cross-sectional area [m^2]
- **P_{core}** — Core pressure [Pa]
- **P_{env}** — Environmental pressure [Pa]
- **L_{fund}** — Fundamental grain length [m] ≈ 10^{-35}?
- **A_1, A_2** — Unified field constants [units vary by normalization]
- **ε_{sat}** — Saturation strain [dimensionless]
- **σ_{max}** — Maximum elastic stress [Pa]

About the Author

Sreeman Mishu Barua is a specialist in technical systems and industrial quality, bringing years' of experience in identifying the hidden variables that govern complex physical structures. His career has been defined by the meticulous inspection of systems where the smallest "unseen" flaw can determine the success or failure of an entire operation.

Rather than approaching the universe through the lens of abstract probability, Barua treats spacetime as a **physical machine**. By applying the same rigorous standards used to optimize modern energy systems and battery ecosystems, he has identified what he perceives as a fundamental "process flaw" in the traditional vacuum-based model of the cosmos.

In his view, modern physics describes a universe that is "failing" its own quality audit. To make the equations of galaxies work, scientists have had to invent "Dark Matter"—a 95% filler that no one can see, touch, or measure. From the perspective of a Lead Inspector, if a process requires that much "missing" material to function, the problem isn't the material; the problem is the **Model of the Process**.

This book is the result of that critical audit. It is a systematic deconstruction of the universe's "unseen" flaws, rewritten through the lens of a professional who refuses to accept "Dark Matter" as a valid excuse for a broken equation. We are going to stop looking for ghosts and start inspecting the machine.

Read more at https://www.linkedin.com/in/s-m-barua/.

www.ingramcontent.com/pod-product-compliance
Lightning Source LLC
LaVergne TN
LVHW090411210926
848480LV00002B/528
* 9 7 9 8 2 3 3 9 3 3 5 0 9 *